Building Information Modeling for a Smart and Sustainable Urban Space

Building Information Modeling for a Smart and Sustainable Urban Space

Edited by

Rafika Hajji
Hassane Jarar Oulidi

iSTE

WILEY

First published 2021 in Great Britain and the United States by ISTE Ltd and John Wiley & Sons, Inc.

ISTE Ltd
27-37 St George's Road
London SW19 4EU
UK

www.iste.co.uk

John Wiley & Sons, Inc.
111 River Street
Hoboken, NJ 07030
USA

www.wiley.com

Library of Congress Control Number: 2021945741

British Library Cataloguing-in-Publication Data
A CIP record for this book is available from the British Library
ISBN 978-1-78630-703-3

MIX
Paper from responsible sources
FSC® C013604

Contents

Preface

Multi-scale Building Information Modeling (BIM) has the potential to become a powerful decision support tool for urban planning. It can be used to facilitate many applications in various fields such as wind simulations, energy studies, noise studies and various types of analyses that require placing a planned architectural design in its context and being able to follow its implementation and evolution in an urban context, and connecting it with other urban components such as networks, road infrastructures, etc. The 3D urban model has become an essential support for better communication with all stakeholders on urban issues.

This book presents the theoretical and practical basis for implementing a multi-scale BIM. It addresses the issues of data acquisition, modeling, integration and information sharing in an interoperable framework. The book analyzes and provides the basics of BIM and Geographic Information System (GIS) integration in the context of urban management, a preliminary step to achieve the intelligent and sustainable management of an urban space. This book also presents practical case studies illustrating some aspects of using the concept of multi-scale BIM to address certain urban issues, including the segmentation of Light Detection and Ranging (LiDAR) data for BIM modeling, the integration of BIM and 3D GIS for property value simulations and the contribution of BIM and 3D GIS to urban renewal.

Rafika HAJJI
Hassane JARAR OULIDI
September 2021

Acknowledgments

I would like to thank all the people who influenced my professional journey and contributed to the evolution of my career, in particular the direction and professors of IAV Hassan II.

My most sincere thanks go to Professor Hassane Jarar Oulidi with whom I collaborated in the writing of this book. My thanks also go to all the doctoral students who have devoted their time and efforts to advance our research questions, especially Ms. Siham El Yamani, Ms. Oumaima Moufid and Mr. Zouhair Ballouch, who have contributed to the writing of this book.

I would also like to thank the ISTE Ltd team for their high-quality professional work.

I will be sure to thank my mother and father for their unwavering support, trust and guidance.

I dedicate this work especially to my husband for his support and encouragement. I thank him for the sacrifices he has made for me. Thank you for the joy and happiness you bring to my life.

Rafika HAJJI
September 2021

I would like to thank the staff and professors at the Hassania School of Public Works (EHTP), in particular the Department of Mathematics, Computer Science and Geomatics.

Special thanks go to my wife Amina for her unfailing support and encouragement.

Hassane JARAR OULIDI
September 2021

List of Acronyms

ADE	Application Domain Extension
AEC	Architecture/Engineering/Construction
AIA	American Institute of Architects
AR	Augmented Reality
B-Rep	Boundary Representation
BDS	Building Description System
BIM	Building Information Modeling
BMLS	Backpack-Mounted Laser Scanners
bSI	buildingSMART International
CAD	Computer-Aided Design
CIM	City Information Modeling
CNN	Convolutional Neuron Network
CPM	Critical Path Method
CSG	Constructive Solid Geometry

DARCES	Data-Aligned Rigidity-Constrained Exhaustive Search
DBMS	DataBase Management System
DEM	Digital Elevation Model
DL	Deep Learning
DM	Dense Matching
DSM	Digital Surface Model
DTM	Digital Terrain Model
ETL	Extract, Transform, Load
FME	Feature Management Engine
FoV	Field of View
GIS	Geographic Information System
GLIDE	Graphical Language for Interactive DEsign
GLoD	Geometric Level of Detail
GNSS	Global Navigation Satellite System
GPS	Global Positioning System
HBIM	Historic Building Information Modeling
HMLS	Hand-held Mobile Laser Scanner
IAI	International Alliance for Interoperability
ICP	Iterative Closest Point
IDBE	Integrated Digital Built Environment
IDM	Information Delivery Manual

IFC	Industry Foundation Classes
IFD	Information Framework for Dictionaries
IMU	Inertial Measurement Unit
ISO	International Organization for Standardization
ISPRS	International Society for Photogrammetry and Remote Sensing
KML	Keyhole Markup Language
LiDAR	Light Detection And Ranging
LoA	Level of Accuracy
LoB	Line of Balance
LoD	"Level of Detail"/"Level of Development"
LoGeoRef	Level of GeoReferencing
LoI	Level of Information
ML	Machine Learning
MMS	Mobile Mapping System
MNO	Modifiable Nested Octree
MVD	Model View Definition
MVS	Multiple View Stereovision
NIBS	National Institute of Building Science
NURBS	Non-Uniform Rational Basic Spline
OGC	Open Geospatial Consortium
RANSAC	RANdom SAmple Consensus

RDF	Resource Description Framework
RGB-D	Red–Green–Blue-Depth
SFM	Structure From Motion
SIFT	Scale-Invariant Feature Transform
SIRS	Spatially Referenced Information Systems
SLAM	Simultaneous Localization And Mapping
SLoD	Semantic Level of Detail
SPR	Spatial Partitioning Representation
SURF	Speeded-Up Robust Features
UBM	Unified Building Model
VLL	Vertical Locus Line
xBIM	eXtensible Building Information Modeling

Introduction

General context

BIM and 3D GIS for a multi-scale modeling of urban space

Urban space is both rich and complex. Its modeling must support the management of this urban complexity through the development of geometrically and semantically rich 3D models. Whatever the use, the availability of a 3D urban model is commonly accepted as a crucial need that requires organizations producing reference data to direct their developments towards the acquisition of 3D geographic reference systems.

The theme "Building" represents a reference component for spatial data infrastructures that shares relationships with other urban objects like infrastructures, city facilities, cadastral parcels, etc. In the Infrastructure for Spatial Information in the European Community (INSPIRE) directive, the theme "Building" is part of the reference data that is required in the European Data Infrastructure. According to the INSPIRE directive, the definition of a building is very broad:

> A building is an enclosed construction above and/or underground, used or intended for the shelter of humans, animals or things or for the production of economic goods. A building refers to any structure permanently constructed or erected on its site.

Introduction written by Rafika HAJJI and Hassane JARAR OULIDI.

Within an urban diversity, the building is an object of reference that is at the center of several issues and is the subject of several studies. The building is, among other things, a space for living, working and human activities, which consumes resources and which defines and controls the dynamics of the urban space.

Faced with a wide range of applications and user needs that are both varied and evolving, the consistency of a general purpose 3D model is difficult to implement in terms of the types of objects to be represented as well as their geometric and semantic accuracy specifications, since the levels of detail and accuracy are strongly related to one's interest in the 3D objects to be represented in a specific application.

The process of digitizing a building leads to a complete, geometrically reliable and precise 3D representation, semantically annotated in the form of a building information system, commonly called BIM (Building Information Modeling). Through a collaborative process, BIM refocuses practices around a highly detailed digital model containing qualitative and quantitative information on a building, and allowing all stakeholders (project owners, architects, engineers, operators, owners, etc.) to coordinate their contributions throughout the lifecycle of the project. This process allows us to optimize the methods of the construction, management and operation of buildings and to become more efficient in terms of cost and time required for the development of a project.

BIM has its origins in 1962, where the basic premises were first established by Douglas C. Engelbart, who in his article "*Augmenting Human Intellect*" described the way in which the architect can perceive the evolution of their project with adjustments to information flows through an object-oriented design. In 1975, Eastman established the link between the architectural design of a building and the field of computer science, and then developed and implemented a Building Description System (BDS) that establishes the basis of object modeling through a model that encapsulates different information and their management within the BDS (Eastman 1975), in which the "element" is the basic unit to which information is added.

Today, BIM is one of the major technological innovations in the field of construction, providing a 3D information base for studies and simulations on an urban scale, including environmental studies, energy, noise, property value simulations, etc. BIM allows the characterization of the geometry of buildings, spatial relationships, quantities as well as properties of construction elements, cost, materials, etc. Beyond its contribution to the economy of construction through cost and time optimization, BIM responds to sustainable development through a model that integrates information on the elements of the model for intelligent and sustainable construction. Upon delivery of the building, an "as-built" version of the BIM represents the actual state of the building, which helps facility managers to undertake maintenance and intelligent building management operations.

At an urban scale, GIS (Geographic Information System) is proving to be a powerful system for managing spatial phenomena. It must therefore be aligned with the increased need for rich and well-structured three-dimensional data that can offer advanced functionalities in a 3D space. The problem is not reduced to a simple extension of 2D GIS solutions by adding a third dimension, but requires consistent modeling, representation, storage and 3D spatial analysis for an optimal management of 3D data.

Through the capture, modeling, storage, manipulation, analysis, sharing and representation of geographically referenced data, 3D GIS describes information about the environment as it is captured at different times in a 3D environment. However, it provides access to data that is less detailed than data from BIM, but more up-to-date and covers a wide spatial extent (Worboys and Duckham 2004), therefore giving BIM and 3D GIS two different scales of modeling and analysis.

Trivially, the development of a 3D GIS is motivated by the increased demand for 3D information, and also by the technological revolution in 3D data acquisition, 3D reconstruction and modeling, new 3D visualization techniques such as virtual, augmented or mixed reality, and 3D spatial analysis. The challenge today is to choose the most appropriate technique for modeling a given spatial problem, from a range of 3D acquisition solutions, which is continuously developing and increasingly accessible to (initially) non-expert users. The difficulty lies rather in the implementation of solutions for processing, optimized storage and knowledge extraction from a 3D dataset. Moreover, the quality and integrity of the acquired data are two important parameters to be taken into consideration in the development of 3D models.

BIM-3D GIS integration: A new paradigm for a smart and sustainable urban space

The planning and management of the built environment requires at least two levels of analysis and planning, either at the city or neighborhood scale (GIS) or at the building scale (BIM). An integration of both BIM and 3D GIS models will be beneficial to adapt urban territories to the digital age. The current research trend is towards the integration of approaches from the geographic information domain (3D GIS) and the architectural and engineering domain (BIM). The challenge is to make a multi-scale modeling of urban space.

The result of this integration is GeoBIM, a hybrid process that combines information from the BIM micro-scale (building) and the GIS macro-scale (neighborhood, city, region, etc.). Thanks to its very detailed and precise information on the elements of a building, the BIM feeds the information represented by the GIS; the latter contains more general information and extends to a wider spatial context.

One example of the potential of 3D GIS is its ability to provide a platform for the simulation of urban issues related to the concept of a "Smart City". If the major issue for politicians today is sustainable construction and the implementation of green strategies for new cities, the upgrading of existing buildings to meet the axes of sustainable development is not to be overlooked. 3D GIS plays a major role in this context. On the other hand, BIM provides very detailed and well-structured information about the building which allows its design, construction, management and operation to proceed in a sustainable and intelligent way.

Faced with a sustained urban dynamic and taking into account economic, social and environmental changes as well as the reforms undertaken, the territories of tomorrow are called upon to develop a capacity for resilience and sustainability capable of meeting the major challenges they face. Thus, the implementation of a new model of development and urban planning that responds to the many challenges of competitiveness, social cohesion, preservation of resources and sustainable development and innovation requires the development of concepts and new approaches to planning and a better integration of opportunities offered by new 3D technologies. The integration of 3D GIS and BIM allows the study of the dynamic relationship between physical and environmental conditions, urban geometry and the properties of each building. Such integration will help achieve smart, resilient and sustainable urban spaces (Niu *et al.* 2015). Having multi-scale urban models not only helps to meet the current requirements for urban space analysis and management, but also helps meet the future and prospective needs of tomorrow's cities.

However, there are several conceptual and technical complexities that arise from BIM-3D GIS integration. This is mainly due to dissimilarities between the two domains in terms of spatial scale, level of granularity and detail (LoD), geometry representation methods, storage and access, and semantic dissimilarities. There are three main levels of 3D BIM-GIS integration: data-level integration, application-level integration and model-level integration. This last level is more flexible until one of the two models (BIM or 3D GIS) is extended through its standard to integrate the data and elements of the other model. Another more advanced level of extension is the development of a meta-model that mediates between the two models at a higher conceptual level. In the literature, the contributions in the integration of BIM and 3D GIS are notable, but are far from being able to solve all the technical problems inherent to this integration (Biljecki and Tauscher 2019). This is a niche area of research that is still active.

Interoperability: a major challenge for multi-scale BIM

In the general context of "Data Sciences", the exchange and sharing of data is unavoidable. Given the heterogeneity of systems, tools and formats, interoperability is recognized as a major challenge in the integration of multi-source data. Interoperability is the ability to ensure that data generated by one user can be correctly interpreted by all other users (Shen *et al.* 2010). Data interoperability enables reliable and efficient information exchange: it is a prerequisite for effective system integration in a collaborative context. The goal is to eliminate or reduce time-consuming and error-prone manual interventions inherent in the operation or exchange of data between software and users.

Interoperability in 3D urban modeling is the fundamental objective behind the development of the CityGML standard. CityGML is an open data model for the storage and exchange of 3D city models. The objective of the development of CityGML is to achieve a common definition of the basic entities, attributes and relationships of a 3D city model. CityGML is based on a very rich geometric, topological and semantic data model. It also provides a multi-scale representation through a number of defined levels of detail for urban objects (Gröger *et al.* 2012). The most commonly used level of detail is that of buildings.

In the field of construction engineering, IFC (Industry Foundation Classes) has been proposed as a standard based on an open object-oriented model. IFC is designed to model the objects of a BIM in terms of geometric information, semantics and relationships between elements to facilitate the interdisciplinary coordination of BIMs, the sharing and exchange of data between IFC-compliant applications, and the transfer and reuse of data in different contexts.

Over the last few years, a number of studies have been conducted to examine the potential for the integration of GIS and BIM (GeoBIM) through the standardization of exchanges between IFC and CityGML. However, the technical issues inherent to the management of interoperability between the two schemas are far from being mastered and still mobilize several research groups such as the EuroSDR GeoBIM working group (Noardo *et al.* 2020).

Scientific contribution to this book

This book deals with important issues around the concept of "multi-scale BIM" from the perspective of intelligent and sustainable urban management. Its main contribution lies in appreciating the potential of multi-scale BIM in the development of a real territorial strategy, allowing professionals and researchers to place territorial data in contexts of analysis and interpretation in order to provide integrated urban knowledge for intelligent territorial management. The concepts treated and the practical use cases presented in this book will provide essential building blocks for the development and implementation of smart cities. The book enables:

– reviewing the state of the art in terms of technologies, methods and approaches to implement a multi-scale BIM;

– addressing the conceptual and technical aspects around the interoperability of BIM and 3D GIS models and their integration under the GeoBIM concept;

– helping professionals in the field of geospatial sciences, architecture and construction to share a common vision of BIM and CIM (City Information Modeling) and to understand their integration;

– presenting three practical case studies illustrating concrete examples of multi-scale BIM implementation.

Book structure

In order to pose the problem in a structured way, this book is divided into eight chapters:

Chapter 1 presents an overview of the basic concepts of BIM technology. It provides fundamental notions related to the concept of BIM with regard to conceptual and technical dimensions.

Chapter 2 presents the different methods for collecting and acquiring 3D data for producing a digital model. This chapter provides an overview of the most common 3D acquisition techniques and tools, with an extended reflection on 3D surveys of urban environments in general, prior to the production of urban models. This chapter also proposes a discussion around the choice of techniques and methods and the integration of multiple and heterogeneous data sources for the 3D reality capture of urban space.

Chapter 3 addresses the modeling methods for the development of a multi-scale BIM. This chapter presents the basis for the production of a digital 3D mock-up and the main approaches developed for 3D reconstruction of its objects, in particular buildings.

Chapter 4 deals with the interoperability of models and gives a global overview on Open BIM standards offering the possibility for professionals to work in an exchange environment that guarantees the interoperability of software within the framework of the "open" and standardized digital mock-up.

Chapter 5 explains the principle of combining BIM and GIS technologies under the concept of GeoBIM. It highlights the benefits of GeoBIM with respect to several topics and discusses the latest research findings in this area.

This theoretical and technical analysis is supported by three chapters presenting three case studies:

– BIM and 3D GIS integration for real estate valuation (**Chapter 6**);

– semantic segmentation of airborne LiDAR data for urban model development (**Chapter 7**);

– use of BIM models for the renovation of urban spaces (**Chapter 8**).

Finally, the book ends with a general conclusion that summarizes the highlights of the concept of "multi-scale BIM" by posing some perspectives on the subject.

Target audience

Because of its content covering the multi-dimensional aspect of multi-scale modeling of urban spaces, this book is intended for a large community of professionals working in the field of land use planning, urbanism, architecture, civil engineering, topographic engineering and geomatics, as well as researchers, academics and doctoral students wishing to specialize in the field of BIM applied to territorial and urban intelligence.

References

Biljecki, F. and Tauscher, H. (2019). Quality of BIM–GIS conversion. *ISPRS Annals of Photogrammetry, Remote Sensing & Spatial Information Sciences*, 4.

Eastman, C.M. (1975). The use of computer instead of drawings in building design [Online]. Available at: https://www.researchgate.net/publication/234643558_.

Gröger, G., Kolbe, T.H., Nagel, C., Häfele, K.H. (2012). OGC City Geography Markup Language (CityGML) Encoding Standard. Document, Open Geospatial Consortium, Wayland, MA.

Niu, S., Pan, W., Zhao, Y. (2015). A BIM-GIS integrated web-based visualization system for low energy building design. *Procedia Engineering*, 121, 2184–2192.

Noardo, F., Harrie, L., Arroyo Ohori, K., Biljecki, F., Ellul, C., Krijnen, T., Eriksson, H., Guler, D., Hintz, D., Jadidi, M.A., Pla, M., Sanchez, S., Soini, V.-P., Stouffs, R., Tekavec, J., Stoter, J. (2020). Tools for BIM-GIS integration (IFC georeferencing and conversions): Results from the GeoBIM benchmark 2019. *ISPRS International Journal of Geo-Information*, 9(9), 502.

Shen, W., Hao, Q., Mak, H., Neelamkavil, J., Xie, H., Dickinson, J., Thomas, R., Pardasani, A., Xue, H. (2010). Systems integration and collaboration in architecture, engineering, construction, and facilities management: A review. *Advanced Engineering Informatics*, 24(2), 196–207.

Worboys, M.F. and Duckham, M. (2004). *GIS: A Computing Perspective*. CRC Press, Boca Raton, FL.

1

BIM: A New Paradigm

1.1. Introduction

Building Information Modeling (BIM) is a new concept that is revolutionizing business practices in the field of construction engineering, through the introduction of new work methods that save time, optimize costs and increase productivity and quality of services during the various phases of the building lifecycle (design, construction, operation, maintenance, demolition). In addition to providing an intelligent digital model to support a construction project, BIM can be seen as a virtual process that integrates several aspects, disciplines and systems related to a building, allowing several stakeholders (architects, engineers, developers, clients, etc.) to collaborate in an efficient and interoperable way (Kymmell 2008).

In a BIM context, the traditional "design–bid–build" approach, which is based on a sequential process and a bi-directional exchange of files, is abandoned in favor of a collaboration that is established at the beginning of the project, in order to produce a design that will be simulated and optimized in terms of quality, cost and time (Azhar *et al.* 2012). The approach is conditioned by the implementation of a collaboration framework that ensures easy and interoperable access to the digital mock-up by the different stakeholders (Figure 1.1).

This chapter presents the fundamental basis for understanding BIM. It begins with a historical overview of BIM (section 1.2), followed by a definition of the concept in terms of its technological and methodological

---

Chapter written by Rafika HAJJI and Hassane JARAR OULIDI.

dimensions (sections 1.3 and 1.4), and then presents some of the main uses of BIM in the AEC (Architecture and Engineering Construction) domain (sections 1.4 and 1.5). Section 1.6 defines the concepts of level of detail (LOD) and dimensions of BIM, while section 1.7 presents the concepts of BIM maturity and capability.

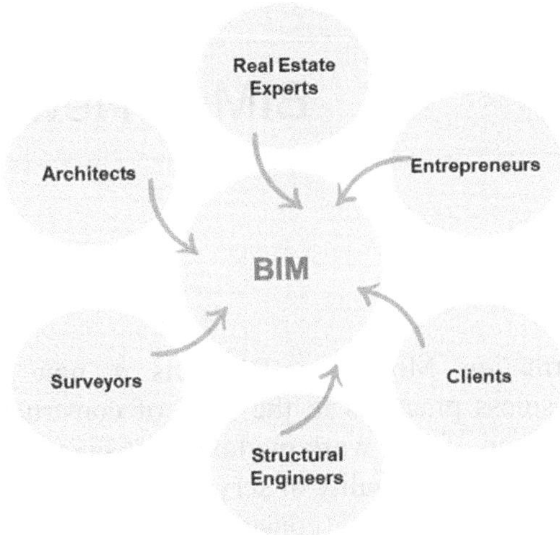

Figure 1.1. *Collaboration of several stakeholders around a BIM model*

1.2. History of BIM

BIM is not new. The basic principles were first established in 1962 by Douglas C. Engelbart, an American engineer, inventor and pioneer of computer science for his work on the development of the human–computer interface. Engelbart described the way in which the architect can perceive the evolution of his project, with adjustments of the information flows through an object-oriented design (Engelbart 1962).

In 1975, Eastman researched the connection between architectural building design and computer science and thus established the premises of the BIM concept. He developed and implemented a Building Description System (BDS) with a group of researchers, which establishes the basis of object modeling through a model that encapsulates different information, in which the "element" is the basic unit to which information is attached. He also described the principles of object-oriented modeling and the notion of producing different design variants and managing them within the BDS (Eastman 1975).

The GLIDE (Graphical Language for Interactive DEsign) project, conducted at Carnegie Mellon University, represents an early CAD (Computer-Aided Design) development for buildings that incorporated parametric modeling (Eastman and Henrion 1977).

From 1984, software development was initiated by the appearance of the first BIM software, ArchiCAD (Radar version). Afterward, other competing solutions such as Revit (Autodesk) or the BIM Digital Project suite followed (Joblot 2018). The International Alliance for Interoperability (IAI) (which later became buildingSMART) was founded in 1995. It is an association of construction companies and software publishers that has developed standards for BIM interoperability, the best known of which is the IFC (Industry Foundation Classes) format.

1.3. BIM: A meta concept

BIM is a new business paradigm in the construction field for which several definitions have been proposed by the scientific community. Among these definitions, the one adopted by the NBIS (The National Institute of Building Science) defines Building Information Modeling (BIM) as "*a computable representation of the physical and functional characteristics of a facility and its related project/lifecycle information, using open industry standards to inform business decision making for realizing better value*". According to this definition, BIM is a digital representation of the physical and functional characteristics of a building and its related project/lifecycle information, using open formats to support decision-making processes.

BIM is a meta-concept that encapsulates several definitions. The "M" in the acronym BIM refers to three concepts that contribute, in a complementary way, to the definition of the paradigm that BIM represents: "**Modeling**"–"**Model**"–"**Management**" (Figure 1.2). Indeed, BIM is a **management** process of the phases of a building lifecycle that is based on the **modeling** of related information through the use of a **digital model** (3D model) that constitutes the basic foundation for the BIM process. The concept thus has a double component: "technological" and "methodological".

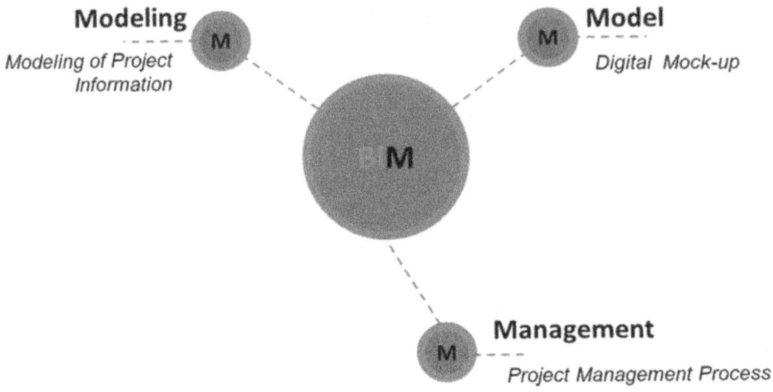

Figure 1.2. *The BIM meta-concept*

The BIM digital mock-up can be considered as a rich database of the building where each element (wall, lamp, opening, heating installations, etc.) is geometrically modeled (3D representation), has alphanumeric attributes (dimensions, brand, type of material, date of installation, state, etc.) and has a behavior with respect to the elements that surround it through relationships with other elements. For example, a door is represented by its material, its dimensions and the relationship it has with the supporting wall. This type of modeling ensures that the components are automatically updated and thus preserves the consistency of the model. This representation complements the purely geometric description of the building's shape produced by CAD (Computer-Aided Design) software.

BIM is based on a collaborative process where several trades (architects, civil engineers, surveyors, structural engineers, etc.) collaborate around a digital BIM model that serves as a central element to support the management processes of a construction project (Pauwels *et al.* 2010). The BIM provides a central resource of diverse knowledge about the building that serves as the basis for making decisions throughout its lifecycle, from design to demolition, including maintenance and operation. Collaboration in a BIM process requires the adoption of standardized processes that define the way information is structured to ensure accessibility, reliability, durability and interoperability of exchanges between the different stakeholders, and during the different phases of a construction project.

1.4. BIM: Between technology and process

With regards to its technological dimension, BIM is based on technical solutions for the conception, design and simulation of projects. A range of BIM tools exist for the production of digital models. These are complemented by other external solutions or integrated into the BIM software environment to undertake several simulations (cost, time, energy, etc.). Beyond the design and construction of the 3D model, the BIM extends to other dimensions, adding other aspects of sustainable development, intelligent management and this, after the design phase.

Today, BIM is integrated into a technological ecosystem using artificial intelligence, digital twin, Big Data, cloud computing, etc. With such integration, the building is thought, designed, built and augmented in its urban environment in order to meet the important challenges of smart cities and sustainable development.

According to its methodological dimension, BIM calls for standardized processes of exchange and structuring of information. Several platforms such as BIMServer are available for sharing and collaboration around the BIM model, where several stakeholders can make updates, notifications and simulations. The process is supervised by a BIM Manager who has control over the model according to the project's BIM agreement and charter. Within a collaborative environment, processes can be developed and integrated to manage the model and continuously update it. BIM is based on a variety of standards including IFC which is the basic standard for BIM.

1.5. BIM in the lifecycle of a building

In a BIM process, the digital model is at the center of the project. This model is unique and models all the geometric and semantic information about the building from design to operation and eventually demolition (Figure 1.3). Thanks to this model, which is constantly updated, coordination and exchanges are improved, errors are detected and resolved before their impact on the cost is induced and simulations are carried out to test the different variants of the project.

The benefits of the BIM approach are projected over the lifecycle of a building which we analyze below.

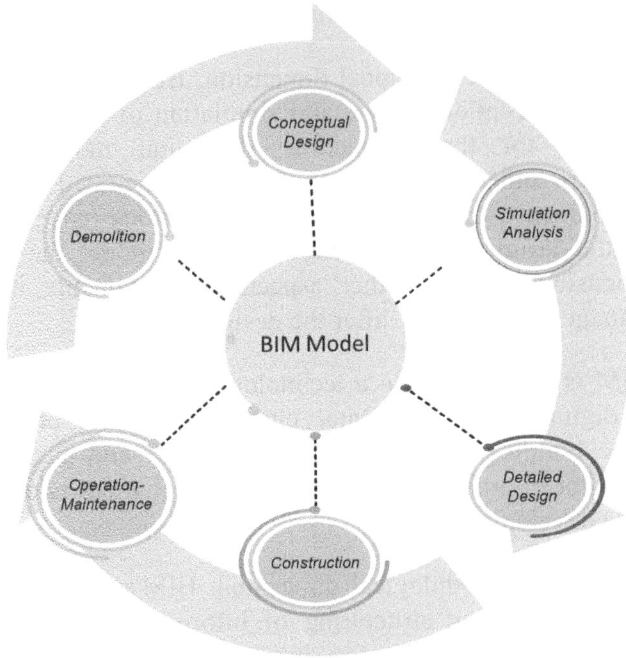

Figure 1.3. *BIM is in the lifecycle of a building. For a color version of this figure, see www.iste.co.uk/hajji/BIM.zip*

1.5.1. *From design to pre-construction*

The different stakeholders of a construction project can benefit from BIM in the design phase. BIM allows for the analysis of different design options by integrating a photographic montage of the project, allowing for photo-realistic visualization and good communication between the different stakeholders. 3D models, animations and virtual tours are tools that help share a co-design of the project between the different stakeholders (Eastman *et al.* 2011).

The ability of BIM to associate various semantic information with the geometric elements of the model allows the designer to integrate technical and performance criteria at a preliminary stage of the project, and to proceed with iterative simulations of the implications of each version of the design. At this level, certain analyses, notably the building's energy performance and cost, as well as structural analyses, are performed to choose between possible scenarios. This approach combines the creativity of the architect-designer with the technological aspects of BIM related to the modeling and structuring of information in an integrated database, providing rich information for testing, adapting and validating the different variants (Eastman *et al.* 2011).

In the pre-construction phase, the BIM model is used to control and simulate the project before its execution in the field. It allows us to make quality and cost estimates. At this stage, an on-site coordination is established to control the execution and establish a site occupation plan, as well as a safety plan (Keymmell 2008). The 3D and 4D dimensions of the BIM for planning and cost estimation, respectively, are used at this level.

1.5.2. *From construction to operation*

This phase concerns the management of the project during construction. Through the BIM model, 4D simulations allow us to monitor the project and ensure the follow-up of its evolution. As the project progresses, the BIM model must be kept up to date to reflect the reality on the ground. Periodic acquisitions are envisaged to recalibrate the model. A "Scan-versus-BIM" process is performed for quality control. At the end of the project, the "as-built" BIM model is the final deliverable that will allow facility managers to prepare monitoring, operation and maintenance plans for the building.

After its construction, the building enters a post-construction phase which assumes that the last version of the model corresponds to the field version. This model enters another lifecycle of the building which concerns the operation and maintenance. Among the applications of BIM in the post-construction phase, we can cite maintenance, risk management and prevention, space planning, real estate evaluation, etc.

1.6. Some transversal uses of BIM

The construction industry is undergoing a double fragmentation. On the one hand, there is a horizontal fragmentation between the different phases of a project: design, planning, conception, construction, etc. On the other hand, there is a vertical fragmentation between the different specialties of project stakeholders: engineers, architects, contractors, etc. (Mazars 2017). BIM offers an effective platform to manage this fragmentation. Its advantages lie in its demonstrated capabilities in the management of construction projects within a framework of coordination and collaboration between different stakeholders and at different phases of the project.

In this section, we present the main benefits of BIM that can occur at different phases. We approach them in terms of a general transversal contribution that can concern several stages of the construction project.

1.6.1. *Visualization*

Visualization is an essential step for a good understanding of the project. Through an interactive visualization support, the communication around the project is easier between the different participants. Visualization is carried out within the software that constitutes the 3D modeling environment of the model, through 3D viewers or augmented reality (AR) applications that allow an increase in the geometric perception of the project, by attaching semantic, attribute or simulated information to the geometric data of the model (Figure 1.4).

Figure 1.4. *Visualization of a BIM model using augmented reality. For a color version of this figure, see www.iste.co.uk/hajji/BIM.zip*

1.6.2. *Coordination and collaboration*

BIM is a process that federates several construction stakeholders around a single digital model. The process established by BIM ensures coordination between different trades: project owners, architects, structural engineers, civil engineers, surveyors, owners, etc. Several times during the design phase, the participants check the coherence between the different themes of the project and collaborate to resolve possible conflicts. This is a crucial stage, where BIM offers its real productivity benefits.

To ensure effective coordination, collaborative platforms such as BIMServer, eXtensible Building Information Modeling (xBIM), BIM 360, Trimble connect, etc., are adopted to share the model at a given time and allow different collaborators to integrate their changes. Some applications such as BIM apps: BIMX®, Bentley Navigator®, Buzzsaw and Autodesk 360®, allow users to share BIM models in a web environment with features such as conflict detection, virtual tours, etc.

1.6.3. *Construction planning: 4D simulation*

The purpose of construction project planning is to coordinate the intervention of the various stakeholders on a construction site, at different stages of the project. Different factors must be taken into consideration to determine optimal space occupancy rates, while avoiding congestion and thus increasing productivity (Francis and Miresco 2016).

Construction planning allows for the planning and coordination of a virtual construction and the monitoring of its progress on the ground. Unlike traditional methods such as Gantt charts that do not allow for the visualization of spatial constraints and resource mobilization by an activity, the BIM model allows for optimal and supervised scheduling through 4D simulation features. This usage calls upon the 4D dimension of BIM, which integrates time into the BIM model by linking the tasks of a construction schedule with the elements of the BIM model via a specialized computer tool offering a 4D interface.

1.6.4. *Cost estimation: 5D simulation*

In the BIM process, the model can automatically provide an indication of the future cost of construction. The object modeling approach adopted by BIM allows us to produce quantity extractions for architectural elements, structural elements and other project components. These quantities can be extracted at different stages of the construction: at the conceptual phase for budget generation and also for financial estimation of changes introduced in the model and invoice verification. Tools for estimating cost from a BIM model include Open BIM Cost estimate, BIM C, BIM Office, BIM Estimate, Vico Office, BIM Vision, Innovaya, JustBIM, Cost X, Cost OS, Dprofiler, Autodesk Quantity Take-off and Autodesk Navisworks.

1.6.5. *Building management and maintenance*

At the end of the construction phase, the building enters an operation and maintenance phase. Research indicates that 85% of the cost of a building's lifecycle is in the post-construction phase and that approximately 10 billion dollars are lost annually in the US in this phase, due to lack of access to information and lack of interoperability (Newton 2004).

The final BIM model represents the as-built version of the project. This model can serve as a reference database to undertake management and maintenance operations through the digital simulation of options and costs. While conventional maintenance, often of an event-driven nature, intervenes on the building without any prior knowledge of the plans and information underlying the walls, such as networks, and can thus induce damage to other parts of the building, BIM-based maintenance allows for the planning of operations while ensuring the integrity and respect of the relationships between the various components and a prior estimate of costs.

The BIM model contains a wealth of information that can be used downstream by managers for better maintenance. For example, the information extracted for a heating/cooling system is location, brand, serial number, operation and maintenance manual, performance data, etc., which will enable the maintenance manager to perform his role effectively (Philips and Azhar 2011).

1.7. BIM dimensions and levels of detail

1.7.1. *The "xDs" of BIM*

One of the potential contributions of BIM is its ability to perform multidimensional simulations that can take place at different phases of the project. These simulations cover several themes in order to optimize the cost and time of the project, while ensuring quality requirements. From 3D, which is geometric modeling, to xD, the BIM model can integrate a multitude of dimensions that can be associated with different simulations.

BIM-4D integrates the notion of time by associating the execution schedules of the different phases of the project (4D = 3D + Time) with the BIM model. The principle of 4D simulation is to link the tasks of a construction schedule with the elements of the BIM model via a specialized computer tool offering a 4D interface. Simulations are used to estimate the time required to execute tasks and thus allow us to control delivery times and evaluate the impact of the introduction of possible changes in the model on the pre-established deadlines.

Two main methods can be used for 4D project modeling: the Critical Path Method (CPM) and the Line Of Balance (LOB) method. CPM is an analysis technique whose goal is to identify the critical path and reference milestones that represent the most important risk in terms of project time lag: each activity is listed with an associated execution time and by specifying its relationships with the other activities. LOB is a graphical planning methodology that, compared to classical methods such as the Gant chart, allows for the following aspects: addition of other features, multiple locations, continuous representation of activities, team planning, integrated visualization and productivity representation (Jongeling and Olofsson 2007).

BIM-5D allows for project cost estimation. The object modeling approach adopted by BIM allows for quantity extractions for architectural elements, structural elements and other project components to be produced at different phases of the project. The financial evaluation of a BIM project can be performed in three options (Eastman *et al.* 2011): 1) an export of quantities to estimating software; 2) a BIM quantity take-off tool; or 3) a direct link between the BIM tool and estimating software.

The cost estimate is strongly related to the quality, geometric accuracy and Level Of Information (LOI) of the BIM model. The LoD specification for building structures generally refers to the LoD (Level Of Development). The LoD 350 is an optimal level for cost estimation (Vitásek and Zak 2018).

BIM 6D allows us to meet the requirements and regulations in terms of sustainable development. In particular, BIM 6D allows the management of the energy performance of a building at different stages of the project. The integration of the 6D dimension of BIM allows us to prepare the model for efficient management in the operation and facility management stage. This dimension allows, for example, easy and precise decision-making on the installation of components to reduce energy consumption.

1.7.2. *Levels of detail*

In a BIM project, the modeling must be designed to meet the predefined objectives for the project. The BIM model must therefore include only the necessary information and elements required for the project. The acronym LoD refers to two different concepts: "**Level of Detail**" and "**Level of Development**". The first concept is associated with the BIM model and the second is related to the BIM process. In general, the level of development refers to the amount of information required for the development of the project, while the level of detail represents the granularity of the information related to the elements of the BIM model.

1.7.2.1. *LoD: Level of Detail*

The concept of level of detail was introduced in 1976 by James H. Clark, who applied it in computer graphics in the context of optimizing image sequences ("Mipmaps") by texture reduction (Figure 1.5): the level of detail of the texture to be used is determined by the size (in pixels) of the object to be mapped. It must be compatible with the image displayed on the screen: when the image size decreases, the texture size decreases accordingly (Clark 1976). The level of detail is thus defined according to the geometric and graphical aspects for optimizing the visualization: adaptation of the level of detail according to the distance from the point of view (Luebke *et al.* 2002). One of the known examples of this method is the Google Earth tool operating with KML (Keyhole Markup Language)/KMZ data (compressed format of KML).

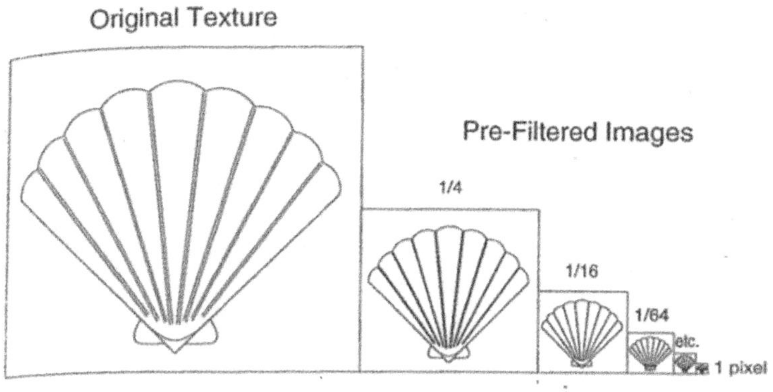

Figure 1.5. *The Level of Detail associated with "Mipmaps" (Clark 1976)*

The concept of LoD was subsequently adopted in 3D modeling of urban objects to address a wide range of applications. The LoD required for a model depends on the needs of the application. This concept integrates the two geometric and semantic aspects of an urban model and represents the level of granularity of the geometric and semantic information of objects. The LoD of an urban object is not only limited to the representation of the complexity of the structure, the richness and quantity of the details of the elements, but also concerns the semantic and topological aspects of the objects (Tang *et al.* 2018).

The most popular LoD classification is the one proposed by CityGML, which is a standard adopted by the OGC (Open Geospatial Consortium) for 3D modeling at the urban scale. CityGML defines five levels of detail for buildings, ranging from the representation of the footprint (LoD0) to the representation of its detailed architecture containing internal elements, such as rooms and equipment (LoD4) (Figure 1.6). Moving from one LoD to a higher level, the representation is refined not only by representing geometric complexity, but also by increasing semantic richness.

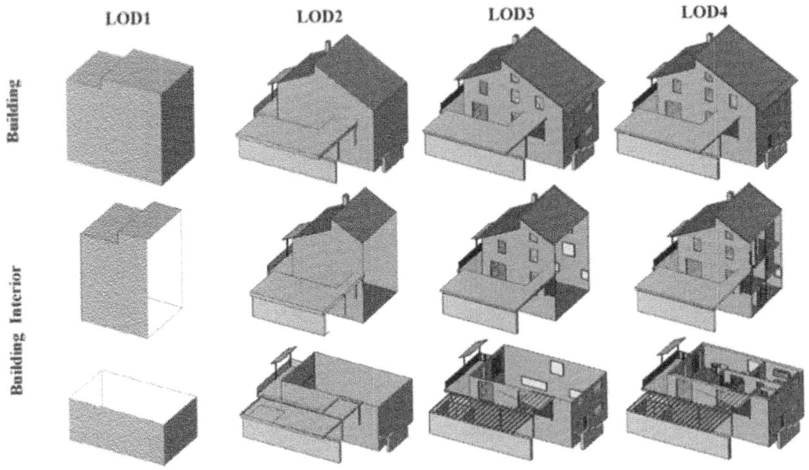

Figure 1.6. *The five levels of detail in CityGML 2.0 (OGC 2012)*

The classification of LoDs by the CityGML standard has been widely adopted, even for applications that do not adopt this standard. Several researches have subsequently proposed extended concepts for CityGML LoDs, such as geometric LoD (GLoD) and semantic LoD (SLoD) (Benner *et al.* 2013).

1.7.2.2. *LoD: Level of Development*

The IFC standard does not integrate the concept of the level of detail as defined in 3D urban models. Multiple geometric representations are associated with an object without being assigned to a defined level of detail.

In the context of BIM, the concept of LoD refers to the level of development of the digital model. The American Institute of Architects (AIA) defines six LoDs for BIM: LoD100, LoD200, LoD 250, LoD 300, LoD 400, LoD 500, ranging from a conceptual representation where elements are represented by symbols or generic descriptions (sketch level) (LoD100) to the "as-built" level (LoD 500). The intermediate levels add precise descriptions (dimensions, thicknesses, orientation, location, etc.) and relationships between elements. Figure 1.7 illustrates the association between the level of development of the model and the project phase.

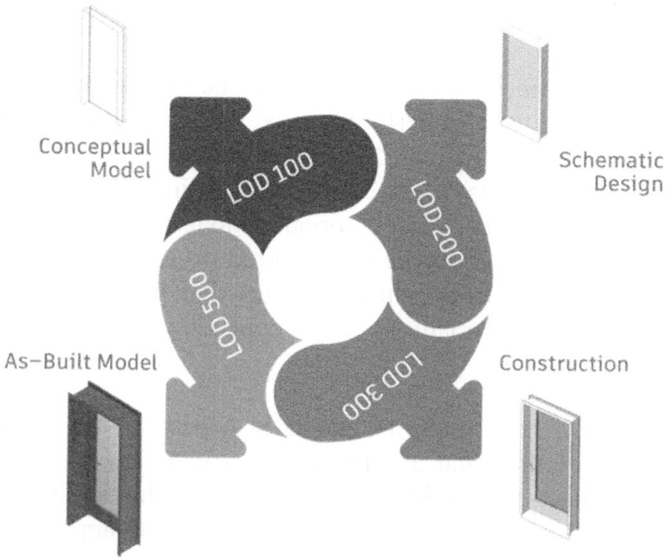

Figure 1.7. *Representation of LoDs according to the stage of the BIM project. For a color version of this figure, see www.iste.co.uk/hajji/BIM.zip*

1.7.2.3. *LOI-LOA-Level of Information Needed*

In addition to the concepts of LoD for "Level of Detail" and LoD for "Level of Development", there are other classification systems such as LoA (Level of Accuracy) or LoI (Level of Information). This is how the term **LoX** came to be used to designate all of these classification systems, which do not only qualify the level of detail or information associated with the objects, but extend to the model as a whole.

The lack of clear standards on the specifications of LoX classifications, the complexity and the multiplicity of definitions for the concept of the level of detail, which can cause ambiguities in the interpretation of the different concepts (LoD–LoI–LoA), led the ISO standard bodies to develop the ISO 19650 standard by adopting the concept of "level of information needed". This concept corresponds to the information required to meet the specific objectives of each information exchange. This would avoid cases of information deficit or excess in the exchanges. This concept is adopted to define the quality, quantity and granularity of information required in an exchange. This information can concern geometric information (level of graphic detail) and alphanumeric information (or LOI) (see: www.iso.org).

1.8. BIM maturity and capability

Succar (2009) defines BIM capability as "the minimum ability required by an organization or team to perform a task, deliver a service, or generate a product". He defines maturity as "the degree of excellence in performing a task". On the other hand, ISO/IEC 33001:2015 defines the maturity level by: "the degree of maturity of an organizational process for a given organizational entity. This degree is measured on an ordinal scale, relative to the Maturity Model used (which explains that the same organizational entity may be at different maturity levels depending on the model chosen)" (Joblot 2018).

The concepts of BIM maturity and BIM capability are closely related. BIM capability refers to the ability of an organization to adopt the BIM process in its projects, while BIM maturity reflects the qualitative nature of BIM ownership. Boton and Kubicki (2014) propose an approach to classify BIM maturity into three levels that are differentiated by the degree of collaboration between different partners (Figure 1.8).

1.8.1. *Level 0: Pre-BIM*

This is the state of the construction industry prior to the implementation of BIM. This is the level of use of CAD models and their exchange through file import and export functions. Each user adopts their own format and drawing, as well as presentation specifications.

1.8.2. *Level 1: Object-oriented modeling*

It is a level in which 2D CAD models are replaced by 3D representations through object-oriented modeling of geometric and semantic aspects. The so-called "BIM" software is adopted for this purpose. In this level, the models and tools remain specific to the organization and the exchanges are done on demand and in a unidirectional way. Parametric modeling is not yet adopted at this stage and exchanges are not yet standardized. This is why the principles of BIM are not yet implemented at this level.

1.8.3. *Level 2: Collaboration based on federated object-oriented models*

In this level, "advanced" modeling is adopted. In addition to the use of CAD and "BIM" software, exchange processes between the stakeholders are established on the basis of well-defined objectives and a connection between the different phases of the project. This level allows the addition of other dimensions to the project such as time (4D) or cost (5D). This level imposes the adoption of an interoperable digital reference model based on the use of standard formats such as IFC, as well as a common definition of the information exchange processes.

1.8.4. *Level 3: Collaboration around an integrated object-oriented model*

In this level, an advanced collaboration is established around a digital model shared between the different stakeholders, who can feed it according to their disciplines. Collaboration is set up in the different phases after the design (co-design and co-modeling). In this case, we are talking about an integrated model that provides several advantages: interoperability of information exchanges, dynamic updating of the model, conflict resolution, coordinated management of activities, etc. This level benefits from the evolution of cloud technology: "Cloud storage" and "Cloud computing", and the development of mobile applications such as augmented reality, as well as 3D printing techniques, which allow a physical restitution of the digital model.

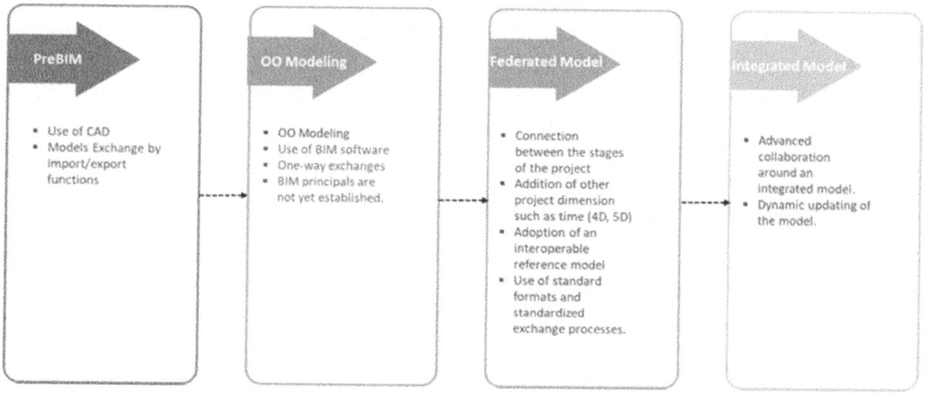

Figure 1.8. *The four levels of maturity of collaborative modeling practices (inspired by Boton and Kubicki (2014))*

1.9. Conclusion

As presented in this chapter, BIM is a meta-concept that introduces a new business paradigm in construction engineering by providing technologies for modeling, simulation, planning and monitoring of construction projects, as well as new business processes based on consultation, communication and information exchange in an interoperable framework.

BIM integrates several dimensions in the management of projects (time, cost, energy, etc.) and thus contributes to the construction of sustainable buildings, at controlled cost and on time. These functions of BIM are ensured through the collaboration between the various stakeholders around a single digital model.

Beyond the use of BIM for the design and construction of new buildings, the case of existing buildings still presents an important challenge for which the BIM approach takes another dimension relating to the survey of the reality through 3D capture systems, and also proposals for scenarios of upgrading the operational aspect of the building, particularly in the case of maintenance and thermal renovation of existing buildings.

1.10. References

Azhar, S., Khalfan, M., Maqsood, T. (2012). Building information modeling (BIM): Now and beyond. *Australasian Journal of Construction Economics and Building*, 12(4), 15-28.

Benner, J., Geiger, A., Gröger, G., Häfele, K.H., Löwner, M.O. (2013). Enhanced LOD concepts for virtual 3D city models. *ISPRS Annals of the Photogrammetry, Remote Sensing and Spatial Information Sciences. Proceedings of the ISPRS 8th 3D GeoInfo Conference & WG II/2 Workshop*, 51–61.

Biljecki, F., Ledoux, H., Stoter, J. (2016). An improved LOD specification for 3D building models. *Computers, Environment and Urban Systems*, 59, 25–37.

Boton, C. and Kubicki, S. (2014). Maturité des pratiques BIM : dimensions de modélisation, pratiques collaboratives et technologies. *SCAN'14, 6ème Séminaire de Conception Architecturale Numérique*, 45–56.

Clark, J.H. (1976). Multiple levels of detail [Online]. Available at: http://people.cs.clemson.edu/~dhouse/courses/405/notes/OpenGL-mipmaps.pdf.

Eastman, C.M. (1975). The use of computer instead of drawings in building design [Online]. Available at: https://www.researchgate.net/publication/234643558_.

Eastman, C.M. and Henrion, M. (1977). Glide: A language for design information systems. *ACM SIGGRAPH Computer Graphics*, 11(2), 24–33.

Eastman, C.M., Eastman, C., Teicholz, P., Sacks, R., Liston, K. (2011). *BIM Handbook: A Guide to Building Information Modeling for Owners, Managers, Designers, Engineers and Contractors*. John Wiley & Sons Inc, Hoboken, NJ.

Engelbart, D.C. (1962). Augmenting human intellect: A conceptual framework. Report, Stanford Research Institute, Menlo Park, CA.

Francis, A. and Miresco, E. (2006). A chronographic method for construction project planning. *Canadian Journal of Civil Engineering*, 33(12), 1547–1557.

Joblot, L. (2018). Contribution à la mise en oeuvre du BIM en rénovation : proposition d'un Modèle de Maturité BIM spécifique. Génie des procédés. *Ecole nationale supérieure d'arts et métiers – ENSAM, 2018*, France.

Jongeling, R. and Olofsson, T. (2007). A method for planning of work-flow by combined use of location-based scheduling and 4D CAD. *Automation in Construction*, 16(2), 189–198.

Kymmell, W. (2008). *Building Information Modeling: Planning and Managing Construction Projects with 4D CAD and Simulations*. McGraw-Hill Education, New York.

Luebke, D., Reddy, M., Cohen, J.D., Varshney, A., Watson, B., Huebner, R. (2002). *Level of Detail for 3D Graphics*. Morgan Kaufmann, Burlington, MA.

Mazars, T. (2017). Exploitation du BIM pour la modélisation chronographique de la planification et la simulation 4D. Thesis, École de technologie supérieure, Montreal.

Newton, R.S. (2004). Inadequate interoperability in construction wastes 415.8 billion. *AEC News*, 13, 342.

OGC (2012). OGC City Geography Markup Language (CityGML) en-coding standard [Online]. Available at: http://web.archive.org/web/20161225015840/ http://www.citygml.org/index.php?id=1522.

Pauwels, P., De Meyer, R., Van Campenhout, J. (2010). Interoperability for the design and construction industry through semantic web technology. In *Semantic Multimedia. SAMT 2010. Lecture Notes in Computer Science*, Declerck, T., Granitzer, M., Grzegorzek, M., Romanelli, M., Rüger, S., Sintek, M. (eds). Springer, Berlin, Heidelberg.

Philips, S. and Azhar, S. (2011). Role of BIM for facility management in academic institutions. *Proceedings of the 6th International Conference on Construction in the 21st Century (CITC-VI),* Kuala Lumpur, Malaysia, July, 5(7), 950–957.

Succar, B. (2009). Building information modelling framework: A research and delivery foundation for industry stakeholders. *Automation in Construction*, 18(3), 357–375.

Tang, L., Li, L., Ying, S., Lei, Y. (2018). A full level-of-detail specification for 3D building models combining indoor and outdoor scenes. *ISPRS International Journal of Geo-Information*, 7(11), 419.

Vitásek, S. and Zak, J. (2018). Cost estimating and building information modelling (BIM) in road construction. *Creative Construction Conference 2018*, 403–410. Budapest University of Technology and Economics.

2

Which Data Sources for the BIM Model?

2.1. Introduction

The three-dimensional modeling of urban spaces is of major importance in many urban applications. Research in this field is in full swing and technological innovations are very notable. Achievements are not limited to the introduction of new techniques for capturing 3D reality, and also tend to change paradigms in many areas developed around geospatial information.

Today, we are witnessing a democratization of access to information, as the techniques available for acquiring 3D data are no longer the prerogative of the expert, but are becoming increasingly accessible to new users, who are initially non-experts. However, the diversity of approaches requires objective choices and a good understanding of the advantages and limitations of each technique.

This chapter focuses on the techniques and approaches for the production of 3D data for a digital model. It proposes a non-exhaustive overview of the most common 3D acquisition techniques and tools with an extended reflection on 3D surveys of urban environments in general, prior to the production of urban models. In this chapter, we also discuss the choice of techniques and methods and the integration of multiple and heterogeneous data sources for the 3D digitization of urban space.

Chapter written by Rafika HAJJI and Hassane JARAR OULIDI.

2.2. Multiple sources for the 3D digitization of urban space

Today, we are witnessing a great technological revolution marked by the development of a wide range of 3D acquisition techniques, allowing us to produce data at different scales: globe, territory, city, neighborhood, building, structural element, etc., for various applications: asset management, urban planning, smart cities, 3D cadastre, building security, real estate evaluation, urban climatology, etc. Compared to traditional methods, the new 3D acquisition techniques are more efficient in terms of speed, accuracy, handling, application and dissemination. The data produced are in image format (remote sensing, photogrammetry) or point cloud format (lasergrammetry, RGB-D (Red-Green-Blue-Depth) images). Within the same technique, several platforms are used (satellite, aerial or terrestrial) according to 3D capture modes that can be either static (terrestrial laser) or dynamic (MMS: Mobile Mapping System, GeoSLAM).

In an urban context, BIM models, whether at the scale of a building or an urban scene (multi-scale BIM model), can be modeled from several sources: lasergrammetric survey, photogrammetric survey or by assembling topographic plans with elevation information (2.5D plans), etc. Laser scanning remains the most used method because it allows the capture of reality quickly and precisely, in the absence of building data.

The multitude of 3D techniques requires the user to make an objective and rational choice in order to adopt the most appropriate method to meet their objectives. Thus, a scientific evaluation in terms of "fitness of use" must scrutinize the options. Often, several techniques can be combined to produce accurate and semantically rich 3D information.

2.2.1. *3D measurement techniques*

Several techniques of 3D measurement have been developed around the principle of imaging sensors and telemetry by optical systems. The types of measurements frequently used are contactless based on the propagation of waves in space (microwaves, light waves, acoustic waves).

There are two main categories of systems: active and passive systems. Passive sensors use ambient light to generate 2D images. These must be associated with depth information to generate a 3D distance by the principle of stereovision triangulation (simultaneous use of a minimum of two sensors) or by the photogrammetric principle based on different shooting positions from a single sensor based on a multi-view approach (Lachat 2019). In active systems, the triangulation principle is extended by using a light source (laser beam, infrared or white light). The triangular configuration consists of the light source, the optical sensor and the object. This is the example of LiDAR (Light Detection And Ranging), which works by emitting laser radiation.

Lasergrammetry and photogrammetry are the most widely used techniques for the 3D acquisition of urban scenes to produce digital models. However, the cost of acquisition and processing by laser scanning is still a major challenge. Recently published research demonstrates the potential of low-cost systems for indoor scene reconstruction. Li *et al.* (2020) proposed to apply RGB-D technology to indoor 3D reconstruction and automatic generation of as-built BIM. The results demonstrate robustness of the method and competitive accuracy with the terrestrial LiDAR system.

The wide deployment of 3D scanning instruments by laser and imaging systems has not replaced tacheometric and GNSS (Global Navigation Satellite System) methods, which are still used for either the acquisition of point data, geometric consolidation or georeferencing of digital models as well as for control and monitoring operations.

2.2.1.1. *Photogrammetry*

Photogrammetry is a science based on the measurement of an object by studying its reproduction in conjunction with photographs or digital images. It is a rigorously accurate technique, where all points captured by the method have a mathematical definition. The 2D digital images are used to derive geometric, radiometric and semantic information about objects in 3D space and to extract colorized point clouds, orthoimages, mesh and textured models, etc. The images can be taken on the ground (terrestrial photogrammetry) or from above (aerial photogrammetry).

Photogrammetry is an old technique, but it still operates in many fields. The convergence between this field of research, which has a well-established scientific basis, and the more recent field of computer vision, has led to the development of fully automatic 3D reconstruction processes from large sets of digital images (Lachat 2019). The automatic and simultaneous determination of the 3D geometry of the scene (the structure), as well as the position of the cameras from image sequences and self-calibration, has become possible through "structure by motion", or Structure From Motion (SFM) algorithms (Snavely *et al.* 2008). We refer the reader to Luhmann *et al.* (2013) for more details on the foundations related to photogrammetry and computer vision.

2.2.1.2. *Lasergrammetry*

Lasergrammetry is a 3D acquisition technique that uses digital sensors or scanners to generate a cloud of known points in three-dimensional coordinates (X, Y, Z) (Figure 2.1). This point cloud is a set of raster 3D points representing the surface measured by the laser scanner. Each point measured with its spherical coordinates is defined in a reference frame attached to the laser scanner. The point cloud is potentially completed by information on the laser intensity feedback and colorimetric information. Some scanners are associated with a video sensor or high-resolution camera to capture images of the object. One of the advantages of the lasergrammetric survey is the speed of acquisition, a homogeneous but relative precision and a high density of points.

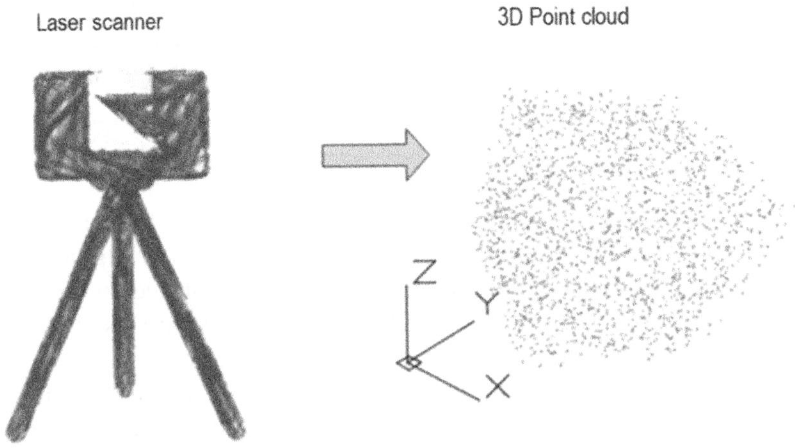

Figure 2.1. *Principle of a 3D scan by lasergrammetry*

In addition to the static laser scanner, the recent development of Mobile Mapping Systems (MMS) offers many advantages in the indoor mapping of complex environments and the production of BIM models. The SLAM (Simultaneous Localization And Mapping) technology behind MMS systems allows for rapid acquisition, automatic real-time registration and access to areas that may be occluded by static scanners. Other mobile solutions have appeared, such as handheld laser scanners and backpack scanners, which bring ergonomics, ease and speed to the scanning process.

With this multiplicity of laser solutions, current research is focusing on the use of point clouds from these techniques in the segmentation and reconstruction processes of BIM models. However, the quality of the generated models is an important issue that, unfortunately, has not yet fully secured its place in research.

2.2.1.3. *RGB-D cameras*

3D data acquisition in the form of RGB-D images has become popular in recent years due to the popularity of RGB-D sensors that provide 2.5D information about the captured 3D object by attaching a depth map with 2D color (RGB) information (Figure 2.2). RGB-D data are simple but effective representations for 3D objects, which can be used for various purposes such as identity recognition, real-time 3D localization, classification, etc. Microsoft started this development in 2010 with the Kinect depth camera, and several other devices followed, such as Intel RealSense, PrimeSense Carmine, Google Tango or the Occipital's Structure Sensor.

These sensors not only have the advantage of low cost, but also capture color and depth images per pixel at adequate resolution and in real time. These features put them ahead of some of the more expensive 3D scanning systems, especially when developing solutions for consumer applications. The number of RGB-D datasets is increasingly available compared to that of other 3D datasets such as point clouds (Cai *et al.* 2017). This data is often used for benchmarks on point cloud segmentation in indoor environments.

The potential of RGB-D sensors is now a subject of research in the production of BIM models. Recent work by Hübner *et al.* (2020) and Li *et al.* (2020) addressed the use of low-cost systems (RGB-D cameras and HoloLens headsets) for 3D reconstruction of indoor scenes.

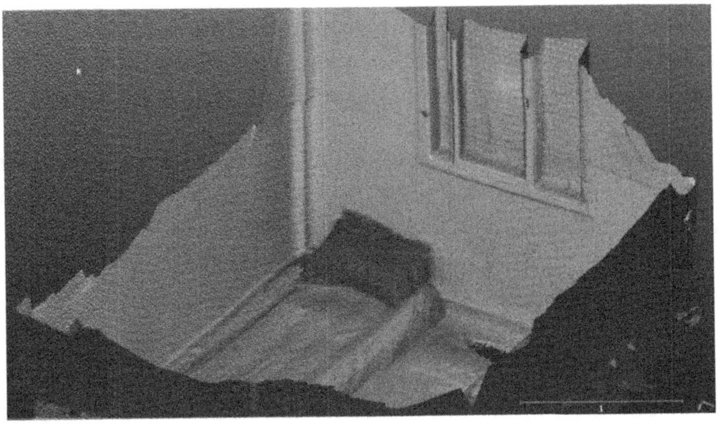

Figure 2.2. *Example of a 3D scene captured in an indoor environment (Cai* et al. *2017). For a color version of this figure, see www.iste.co.uk/hajji/BIM.zip*

2.2.1.4. *Tacheometry and GNSS*

Tacheometric surveys and GNSS surveys are point surveys that collect coordinates of a set of points or feature lines of objects by topometric processes (radiation, bearing, triangulation, etc.) (Figure 2.3). The coordinates obtained are attached to a local coordinate system associated with a projection. Tacheometric surveys are generally used for control operations (monitoring, auscultation), geometric consolidation or georeferencing.

Figure 2.3. *Characteristic line survey by total station. For a color version of this figure, see www.iste.co.uk/hajji/BIM.zip*

Georeferencing of BIM models plays a critical role in BIM and GIS integration. The most accurate method is to georeference the mockups by a surveyor using a total station or GNSS system. The need for accurate lasergrammetric surveys that are geographically referenced has led hardware manufacturers to develop scanners with total station capabilities.

2.2.2. *Selection criteria*

Faced with a wide range of 3D acquisition techniques, choices are made according to several criteria: the envisaged application, the scale of the area to be surveyed, the required precision, the objects to be scanned, etc. Often, the acquisition techniques are not only competitive but complement each other to produce rich 3D data through a process of integration or data fusion.

Photogrammetry and lasergrammetry are the two most widely used techniques for the 3D digitization of urban space. One of the contexts in which their complementarity can be highlighted is the heritage field, where laser scans are complemented by a photogrammetric shooting mission, which allows us to record architectural details and produce a photo-realistic texturing of a heritage building (Figure 2.4).

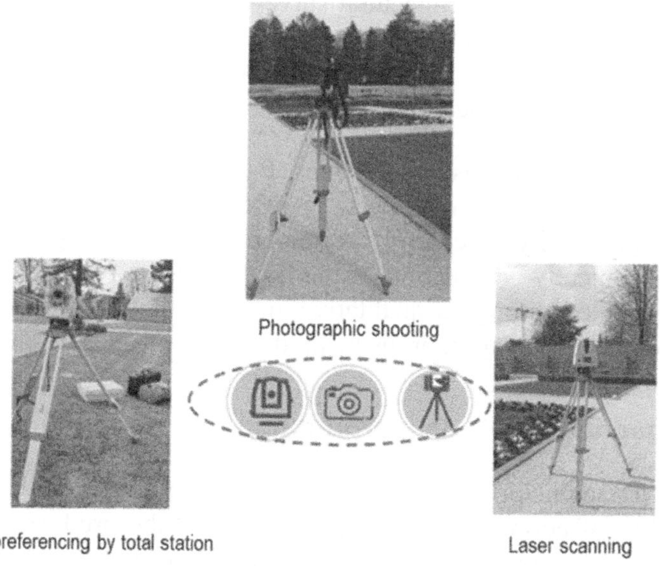

Figure 2.4. *Fusion of laser/photogrammetric data and tacheometric surveys for 3D site acquisition. For a color version of this figure, see www.iste.co.uk/hajji/BIM.zip*

The choice of the type of system (terrestrial, aerial, static, dynamic) can only be made with regard to the specifications. The accuracy required is an important decision-making factor which supposes that, when it comes to making choices, the user should have a quality study that can be used as an objective reference. The quality of the deliverable depends on the knowledge and control of several sources of error: data acquisition errors (instrumental errors and errors due to operating mode), errors related to the characteristics of the object to be measured (geometry, extent, color, types of materials), errors due to the measurement environment (atmospheric conditions), and finally, errors due to the processing chain (Landes 2020).

The extent and complexity of the scene will guide the choice of platform to be used. The acquisition modalities of a system are to be taken into consideration. Indeed, each measurement system has characteristics and specifications of acquisition that must be known, mastered and taken into account in the adopted choice. The type of system (mobile or static), the minimum or maximum acquisition range, the adaptability to the conditions of the site to be surveyed, the type of objects to be acquired, etc., are all parameters that must be considered (Landes 2020).

Static methods (terrestrial photogrammetry or terrestrial LiDAR) still prove their interest in 3D architectural surveys where geometric and textural detail of the object is sought. However, the growing need for urban models has boosted the use of mobile platforms, allowing a faster and more exhaustive survey such as MMS, photogrammetry by drone or airborne LiDAR.

With multi-sensor acquisition systems, the consolidation of clouds from different sources is a crucial need. Software solutions, which do not provide the possibility to control the underlying processing, are far from providing an objective evaluation of the quality of the results. This quality must be controlled through a qualitative and quantitative evaluation of the consolidation results using rigorous methods.

2.3. Approaches for 3D data production

Today, 3D acquisition techniques are undergoing a revolution, offering the possibility of adopting several approaches for the production of urban models. These approaches are grouped into three main classes: (1) point cloud-based approaches; (2) image-based approaches; and (3) hybrid approaches (Figure 2.5).

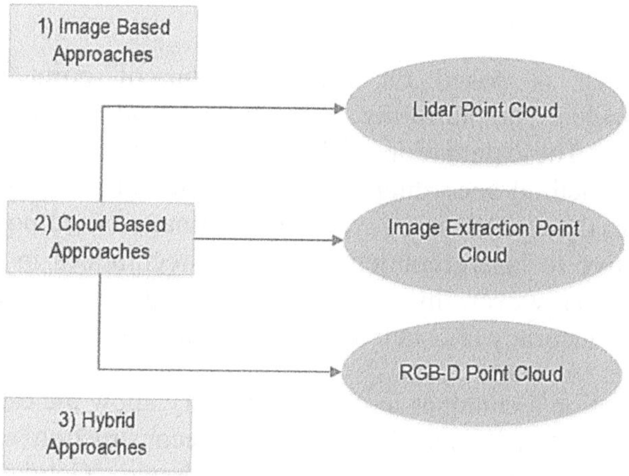

Figure 2.5. *3D acquisition approaches*

2.3.1. *Point cloud-based approaches*

Point clouds can be acquired from different sources (Figure 2.6; Xie *et al.* 2019): (1) image-derived point clouds; (2) LiDAR point clouds; or (3) RGB-D camera-derived point clouds. They are acquired at different scales and can often be combined for accurate 3D reality capture. The choice of techniques depends on the applications and the scale of the scene.

Figure 2.6. *Different point cloud acquisition techniques at different scales. For a color version of this figure, see www.iste.co.uk/hajji/BIM.zip*

2.3.1.1. *Image-derived point clouds*

This approach is based on the extraction of point clouds from stereoscopic images. The platforms are multiple: aerial, satellite or terrestrial photogrammetry. The calculation of 3D points is done via an automatic or semi-automatic method according to the principles of photogrammetry and computer vision theory (Westoby *et al.* 2012). Image-based point clouds are a good alternative to lasergrammetric scanning techniques in terms of cost and accessibility. However, this indirect method requires rigorous processing to achieve good accuracy (Rosnell and Honkavaara 2012).

Image correlation techniques in photogrammetry allow the establishment of Digital Elevation Models (DEM) for the 3D reconstruction of large urban scenes and the generation of dense point clouds (Gehrke 2010). The development of several algorithms such as SFM (Structure From Motion), Dense Matching (DM) and Multiple View Stereovision (MVS) has revolutionized photogrammetry techniques towards automatic and simultaneous multi-view image processing methods and the generation of large volumes of point clouds (Furukawa and Ponce 2010). However, these point clouds do not achieve accuracy comparable to conventional photogrammetry or LiDAR (Xiao *et al.* 2013a).

Cloud extraction is based on matching methods that are classified into two categories: (1) object space-based methods and (2) Image Space-Based (ISB) methods. Object-based methods, also known as Vertical Locus Line (VLL), are reported to be effective methods for point cloud extraction using multiple image matching (Rhee and Kim 2016). ISB correlation techniques use the principle of epi-polar geometry for image matching. These techniques are based on detecting points of interest with associated image descriptors and searching for matches between these image descriptors. SIFT (Scale-Invariant Feature Transform) and SURF (Speeded-Up Robust Features) are operators that have demonstrated their performance in this approach through several applications such as object recognition, 3D scene modeling, video tracking, mapping and robot localization (Lindeberg 2013). These matching approaches are invariable with regards to dissimilarities between images. Their performance has also been proven for drone images (Irschara *et al.* 2010).

2.3.1.2. *LiDAR point clouds*

LiDAR is a currently popular technique for the acquisition of 3D point clouds. With the evolution of technology and thanks to the direct georeferencing integrated in the scanner, the 3D position of the points is deduced by simple geometric construction. With the evolution of airborne LiDAR and mobile LiDAR (MMS), large urban scenes can be acquired quickly and with good resolution.

Several LiDAR technologies have been developed to enable the dynamic acquisition of urban spaces. Notably, Hand-held Mobile Laser Scanner (HMLS) technologies (James and Quinton 2014); Mobile Mapping Systems (MMS) (Thomson *et al.* 2013), Simultaneous Localization And Mapping (SLAM) and recently Backpack-Mounted Laser Scanners (BMLS) (Lauterbach *et al.* 2015) offer fast solutions for 3D capture of complex environments (indoors and outdoors) and generation of massive point clouds. In contrast, software developments have not kept up with this hardware revolution and are still far from being able to offer efficient solutions for processing and exploiting massive point clouds. These solutions are highly impacted by the five V's of the Big Data problem (Velocity, Veracity, Volume, Value, Variety) (Poux 2019).

The advantages of LiDAR in terms of speed, density and accuracy allow the generation of massive point clouds of individual buildings as well as large urban areas. However, these point clouds rarely constitute the final product. They are exploited and processed for the generation of other deliverables such as maps, 3D models and BIM models. Compared to conventional photogrammetry, acquired LiDAR point clouds are more expensive and do not contain spectral information (Xie *et al.* 2019).

2.3.1.3. *RGB-D point clouds*

Over the last decade, RGB-D sensors have attracted a lot of interest thanks to their ability to provide a color image and a depth map simultaneously, and at an affordable cost. These cameras have already proven their worth in robotics and computer vision applications. Recently published research shows their potential in indoor scene reconstruction. For instance, Li *et al.* (2020) proposed to apply RGB-D technology to indoor 3D reconstruction in the context of automatic as-built BIM generation. The results are promising and offer competitive accuracy with the terrestrial LiDAR system.

RGB-D data are also used for the evaluation of some segmentation algorithms or for training deep-learning algorithms. We cite the example of S3DIS (Stanford large-scale 3D Indoor Spaces dataset) which is a large RGB-D dataset captured in urban environments[1].

2.3.2. *Image-based approaches*

In principle, 3D data from photogrammetric sensors are geometrically constructed from at least two images using perspective geometry. The 3D acquisition in image mode is used in many applications: 3D documentation of cultural and archaeological heritage, monitoring and analysis of structural deformations, urban planning, navigation and telecommunications, etc.

The images can be acquired by satellite or by aerial or terrestrial sensors. These images are processed according to the classic photogrammetric pipeline: camera calibration, orientation, shape extraction and orthophoto generation, the basis of which is already established (Kraus and Waldhäusl 1998). A classification of methods can be established based on the number of acquired images and their characteristics by distinguishing between stereo images, monoscopic images and multiple images (Tao 2015). One of the basic applications of photogrammetry is the extraction of the digital surface model from aerial photographs.

Affordability and low altitude acquisition have made UAV (Unmanned Aerial Vehicle) imagery an attractive choice for a number of applications, including those requiring repeatability for monitoring purposes such as crop monitoring, environmental change, archaeological sites, disaster monitoring and urban traffic management. However, the quality of drone images is not comparable to that of aerial images. The main causes are related to the difficulty of calibration, the reduced Field of View (FoV), the low texture and the accuracy of direct georeferencing with the GPS/IMU (Inertial Measurement Unit system) (Rhee and Kim 2016).

2.3.3. *Hybrid approaches*

With the multiplicity of 3D acquisition systems and the development of multi-sensor platforms, coupling between different acquisition techniques is a major issue for a more accurate 3D capture of reality. With the integration

1 http://buildingparser.stanford.edu/dataset.html.

of image sensors, laser scanning and the GNSS/IMU positioning system, the MMS system is a good example of multi-source acquisition of geo-referenced spatial data (Tao 2015), allowing the detailed and accurate 3D capture of large urban scenes.

Most applications related to large and complex sites integrate multiple sensors and techniques for accurate and more complete 3D survey (Remondino 2011) (Figure 2.5). For a long time, several workflows have adopted the integration of images, LiDAR data and GIS maps for the production of 3D city models (Haala and Brenner 1999; Tao 2015). Several integration scenarios are well established in the literature, including the integration of aerial images with airborne LiDAR. Research conducted by Varol *et al.* (2019) integrates and compares airborne LiDAR data and KOMPSAT-3 stereo imagery for illegal construction detection. Indeed, compared to photogrammetry, LiDAR produces a more accurate 3D point cloud. However, occlusion problems give failures at the boundaries of the acquired region. Thus, a combination of point clouds with aerial photos is very popular. Their main advantage lies in the orthorectification or texturing of 3D models (Li and Wu 2008).

The integration of ground and airborne LiDAR has also been investigated in the literature. Kedzierski and Fryskowska (2014) propose a method for integrating and processing data from terrestrial and aerial LiDAR based on signal analysis for identification and data transformations between the two data. Another example of integration focuses on the coupling between terrestrial laser and photogrammetry. This integration is used in heritage applications for the reconstruction of HBIM (Historic BIM) models to provide very rich geometric and semantic 3D models for different uses (restoration, renovation, simulation, etc.).

2.4. Integration of multi-source data

Data from several 3D sensors are by nature heterogeneous. Their exploitation in an integrated context comes up against problems of geometric and semantic mismatch. This heterogeneity can be linked to the types of acquired data (point clouds, images), to differences in dimensions (2D/3D data), in reference frames, in formats, in accuracies, in resolutions (multi-resolution point clouds) and in measurement conditions. The fusion of multi-source data is not a trivial task and requires integration methods whose complexity increases in the presence of a dense and heterogeneous urban environment. Understanding and analyzing the different data sources are a prerequisite to any integration exercise. This allows spatial and temporal inconsistencies to be identified and avoids inconsistent or even erroneous results due to large disparities between sources.

Different data integration scenarios can exist. We cite the example of recalibration of several multi-scale and possibly multi-resolution point clouds. This is the case where detailed 3D surveys on some objects of interest (performed by handheld scanners or 3D cameras) are used to complete the initial point cloud. This example finds a place in heritage literature (Shao *et al.* 2019). Also, DSMs from photogrammetric surveys can also be used to complete the survey of extensive sites, such as in archaeology (Guidi *et al.* 2009). Another example of integration concerns the combination of interior and exterior surveys of buildings. In this context, MMS can be used to acquire external scenes complementary to indoor surveys. In this case, registration is a great challenge as the geometric overlap between the two scans is very limited. To overcome this geometric constraint, Cohen *et al.* (2016) propose a matching method by exploiting semantic information. Another example addressing LiDAR data integration deals with the coupling of ground and airborne lasergrammetric data, for 3D modeling of urban scenes. This issue has been addressed in the literature by several researchers including Kedzierski and Fryskowska (2014), Yang *et al.* (2015) and Zieher *et al.* (2018).

Finally, the registration of photographs on a point cloud is required for certain applications requiring a precise description of the geometry of the objects which cannot be reached by recording devices available in the scanners. This registration of 2D images on a 3D point cloud, called texturing, allows a photorealistic rendering of objects. Also worth mentioning is the work of Gerke and Xiao (2014), who combine airborne laser point clouds with aerial spectral images for improved scene classification.

2.5. General discussion

Today, traditional acquisition methods are impacted by the changes in the field of geospatial information. New 3D acquisition techniques are constantly evolving, offering more efficient 3D capture solutions as well as well-developed processing chains that allow the exploitation of 3D images or point clouds for the derivation of products such as orthophotos, MNT/MNS, virtual tours, 3D models and BIM mock-ups. However, the uses of 3D are essentially technology-based ("by what is technically possible"). The possibilities offered by sensors often direct users towards other prospective uses. Contrary to this reflection, the definition and understanding of needs are a crucial step that must be taken upstream of the 3D project through a specification allowing a rational choice of techniques and processes that are adapted to the required deliverables (Landes 2020).

Beyond the acquisition of data, whose processes are fairly well mastered, the major issue is to do with the processes of their exploitation in specific contexts, their integration and the evaluation of their quality. Indeed, the 3D scanning processes are well established, but their application in certain contexts requires the mastery of the acquisition protocol. For example, different exposure conditions during the scan of an indoor scene require pre-processing to reduce/cancel the impact of overexposure on the quality of the generated model. Also, the evaluation of the quality of models produced by different sensors is a crucial need but does not yet occupy an important part of the research.

Beyond the choice of systems, particular attention must be paid to the sharing and broadcasting format. Indeed, the volume of data generated in the case of large scenes imposes the use of adapted formats allowing us to optimize their display and their management by the user.

3D data is now a growing need for spatial management, but the cost of acquiring, updating and exchanging it limits its appropriation and proper use. Also, 3D models have not yet reached maturity for a good number of organizations that are in the preliminary stages of 3D appropriation.

A 3D urban model must be seen as a basis for the integration of urban knowledge (Falquet and Métral 2006), which demonstrates the usefulness of bringing the issues of collaboration to bear on the development of digital models before their use. Indeed, urban digital models encompass several themes and integrate several stakeholders who must collaborate in the production, updating and management of these digital models. The data must be interoperable and meet the common specifications of the reference model.

The production of digital models at the scale of a territory, their maintenance and updating require significant resources that often exceed budgets. Collaboration around their production is a promising solution to be implemented through two perspectives: (1) the fusion of knowledge distributed between several datasets to produce a 3D reference data and (2) the co-production of a 3D model according to common specifications in order to share costs, on the one hand, and to ensure technical and semantic interoperability, on the other hand (Hajji and Billen 2016).

2.6. Conclusion

Faced with a wide range of 3D data sources, a good understanding of the application needs is crucial for the right choice of an acquisition technique. The user is called upon to make objective choices with regard to the application and the technical specifications in order to avoid cases of under/overproduction of data. Paradoxically to a logic centered on the need which justifies the use of the adapted acquisition technique, the technology can trigger the need and make prospective applications emerge, allowing us to take advantage of the benefits offered by the technique in question.

Urban space is both rich and complex. Its 3D digitization and documentation are at the center of several themes. Over the last decade, the increasing use of laser scanners and photogrammetric reconstruction methods, the development of processing algorithms and the increase in computing power of computers have led to the creation of gigantic point cloud datasets. However, the great challenge lies in the segmentation and structuring of massive 3D data, their optimized storage and their exploitation for knowledge extraction and semantization of urban models.

2.7. References

Cai, Z., Han, J., Liu, L., Shao, L. (2017). RGB-D datasets using microsoft kinect or similar sensors: A survey. *Multimedia Tools and Applications*, 76(3), 4313–4355.

Cohen, A., Schönberger, J.L., Speciale, P., Sattler, T., Frahm, J.M., Pollefeys, M. (2016). Indoor-outdoor 3D reconstruction alignment. In *Computer Vision – ECCV 2016. ECCV 2016. Lecture Notes in Computer Science*, Leibe, B., Matas, J., Sebe, N., Welling, M. (eds). Springer, Cham.

Falquet, G. and Métral, C. (2006). Integrating urban knowledge into 3D city models. *Proceedings of the 1st International Workshop on Next Generation 3D City Models*, European Spatial Data Research, 23–27.

Furukawa, Y. and Ponce, J. (2010). Accurate, dense, and robust multiview stereopsis. *IEEE Transactions on Pattern Analysis and Machine Intelligence*, 32(8), 1362–1376.

Gehrke, S. (2010). Radiometric processing of ADS imagery: Mosaicking of large image blocks. *Proceedings of the ASPRS Annual Conference*, 26–30, San Diego, CA.

Gerke, M. and Xiao, J. (2014). Fusion of airborne laserscanning point clouds and images for supervised and unsupervised scene classification. *ISPRS Journal of Photogrammetry and Remote Sensing*, 87, 78–92.

Guidi, G., Remondino, F., Russo, M., Menna, F., Rizzi, A., Ercoli, S. (2009). A multiresolution methodology for the 3D modeling of large and complex archeological areas. *International Journal of Architectural Computing*, 7(1), 39–55.

Haala, N. and Brenner, C. (1999). Virtual city models from Laser altimeter and 2D map data. *Photogrammetric Engineering and Remote Sensing*, 65, 787–795.

Hajji, R. and Billen, R. (2016). Collaborative 3D modeling: Conceptual and technical issues. *International Journal of 3-D Information Modeling (IJ3DIM)*, 5(3), 47–67.

Hübner, P., Clintworth, K., Liu, Q., Weinmann, M., Wursthorn, S. (2020). Evaluation of HoloLens tracking and depth sensing for indoor mapping applications. *Sensors*, 20(4), 1021.

Irschara, A., Kaufmann, V., Klopschitz, M., Bischof, H., Leberl, F. (2010). Towards fully automatic photogrammetric reconstruction using digital images taken from UAVs. In *ISPRS TC VII Symposium – 100 Years ISPRS*, vol. XXXVIII, Part 7A, Wagner W. and Székely, B. (eds), Vienna, Austria, July 5–7.

James, M.R. and Quinton, J.N. (2014). Ultra-rapid topographic surveying for complex environments: The hand-held mobile laser scanner (HMLS). *Earth Surface Processes and Landforms*, 39, 138–142.

Kedzierski, M. and Fryskowska, A. (2014). Terrestrial and aerial laser scanning data integration using wavelet analysis for the purpose of 3D building modeling. *Sensors*, 14(7), 12070–12092.

Kraus, K. and Waldhäusl, P. (1998). *Manuel de photogrammétrie : principes et procédés fondamentaux*. Hermes, Paris.

Lachat, E. (2019). Relevé et consolidation de nuages de points issus de multiples capteurs pour la numérisation 3D du patrimoine. Thesis, INSA, Strasbourg.

Landes, T. (2020). Contribution à la segmentation et à la modélisation 3D du milieu urbain à partir de nuages de points. Doctoral dissertation, Université de Strasbourg.

Lauterbach, H., Borrmann, D., Heß, R., Eck, D., Schilling, K., Nüchter, A. (2015). Evaluation of a backpack-mounted 3D mobile scanning system. *Remote Sensing*, 7, 13753–13781.

Li, Y. and Wu, H. (2008). Adaptive building edge detection by combining LiDAR data and aerial images. *The International Archives of the Photogrammetry, Remote Sensing and Spatial Information Sciences*, 37(Part B1), 197–202.

Li, L., Yang, F., Zhu, H., Li, D., Li, Y., Tang, L. (2017). An improved ransac for 3D point cloud plane segmentation based on normal distribution transformation cells. *Remote Sensing*, 9(5), 433.

Li, Y., Li, W., Tang, S., Darwish, W., Hu, Y., Chen, W. (2020). Automatic indoor as-built building information models generation by using low-cost RGB-D sensors [Online]. Available at: www.mdpi.com/1424-8220/20/1/293/htm.

Lindeberg, T. (2013). Image matching using generalized scale-space interest points. *International Conference on Scale Space and Variational Methods in Computer Vision*. Springer, Berlin, Heidelberg.

Luhmann, T., Robson, S., Kyle, S., Boehm, J. (2013). *Close-range Photogrammetry and 3D Imaging*, 2nd edition. De Gruyter, Berlin.

Poux, F. (2019). The Smart Point Cloud: Structuring 3D intelligent point data. Doctoral Dissertation, University of Liège.

Remondino, F. (2011). Heritage recording and 3D modeling with photogrammetry and 3D scanning. *Remote Sensing*, 3(6), 1104–1138.

Rhee, S. and Kim, T. (2016). Dense 3D point cloud generation from UAV images from image matching and global optimazation. *The International Archives of Photogrammetry, Remote Sensing and Spatial Information Sciences*, 41, 1005.

Rosnell, T. and Honkavaara, E. (2012). Point cloud generation from aerial image data acquired by a quadrocopter type micro unmanned aerial vehicle and a digital still camera. *Sensors*, 12(1), 453–480.

Shao, J., Zhang, W., Mellado, N., Grussenmeyer, P., Li, R., Chen, Y., Wan, P., Zhang, X., Cai, S. (2019). Automated markerless registration of point clouds from TLS and structured light scanner for heritage documentation. *Journal of Cultural Heritage*, 35, 16–24.

Snavely, N., Seitz, S.M., Szeliski, R. (2008). Modeling the world from internet photo collections. *International Journal of Computer Vision*, 80(2), 189–210.

Tao, J. (2015). Combination of LiDAR and SAR data with simulation techniques for image interpretation and change detection in complex urban scenarios. Doctoral dissertation, TU München, Munich.

Thomson, C., Apostolopoulos, G., Backes, D., Boehm, J. (2013). Mobile laser scanning for indoor modelling. *ISPRS Annals of Photogrammetry, Remote Sensing and Spatial Information Sciences*, II-5/W2, 289–293.

Varol, B., Yılmaz, E.Ö., Maktav, D., Bayburt, S., Gürdal, S. (2019). Detection of illegal constructions in urban cities: Comparing LiDAR data and stereo KOMPSAT-3 images with development plans. *European Journal of Remote Sensing*, 52(1), 335–344.

Westoby, M.J., Brasington, J., Glasser, N.F., Hambrey, M.J., Reynolds, J.M. (2012). Structure-from-motion photogrammetry: A low-cost, effective tool for geoscience applications. *Geomorphology*, 179, 300–314.

Xiao, J., Owens, A., Torralba, A. (2013a). Sun3d: A database of big spaces reconstructed using SfM and object labels. *Proceedings of the IEEE International Conference on Computer Vision*, 1625–1632.

Xiao, J., Zhang, J., Adler, B., Zhang, H., Zhang, J. (2013b). Three dimensional point cloud plane segmentation in both structured and unstructured environments. *Robotics and Autonomous Systems*, 61(12), 1641–1652.

Xie, Y., Tian, J., Zhu, X.X. (2019). A review of point cloud semantic segmentation. arXiv preprint arXiv:1908.08854.

Yang, B., Zang, Y., Dong, Z., Huang, R. (2015). An automated method to register airborne and terrestrial laser scanning point clouds. *ISPRS Journal of Photogrammetry and Remote Sensing*, 109, 62–76.

Zieher, T., Toschi, I., Remondino, F., Rutzinger, M., Kofler, C., Mejia-Aguillar, A., Schlögel, R. (2018). Sensor- and scene-guided integration of TLS and photogrammetric point clouds for landslide monitoring. *ISPRS*, 422, 1243–1250.

3

Development of the BIM Model

3.1. Introduction

Several fields such as indoor navigation, cadastre, construction engineering, building management and energy efficiency require accurate and detailed 3D models of buildings. Other applications require, in addition, the consideration of the interactions between the building and its geographical environment. This is the case for real estate expertise, urban management and the broader concept of smart cities. Between these applications, the differences lie in the requirements in geometric and semantic precision and in the level of detail of the digital models.

With the concept of smart cities and the multiplicity of sustainable development applications, urban digital models must be rich enough to encapsulate several uses and respond to several urban issues. In addition to geometry, the semantics of the models are of major importance, allowing them to be used as real tools for urban management.

Many 3D modeling approaches have been proposed during the past few decades. However, despite the progress made in computer vision, photogrammetry and LiDAR technologies, the automatic reconstruction of 3D urban scenes is still a very active field of research, and some technical challenges still need to be solved.

This chapter presents the basis for the production of a BIM model and the main approaches developed for the 3D reconstruction of its objects, in particular, buildings.

Chapter written by Rafika HAJJI and Hassane JARAR OULIDI.

3.2. Issues around 3D urban models

In the field of Earth sciences, a "geospatial" model is used to refer to models attached to the Earth's surface. In its digital version, a model is a structure assembling several types of data: geometric and semantic that can be maintained and managed by a computer to perform a number of actions that can benefit from machine intelligence. A model does not only contain a collection of data and information, but also includes a number of implicit and/or explicit relationships, allowing the management of interactions between objects in a coherent and integrated way.

3D urban models are designed to respond to urban issues and help the various stakeholders in the decision-making process. These models are supposed to respond to several uses without representing the whole reality. The quality of a 3D model is hence evaluated in terms of its purpose and the forms of its use (Abdul-Rahman and Pilouk 2007).

Beyond the visualization aspect, urban models cover a number of uses and must be designed for intelligent management of the city and implementation of sustainable development actions, hence the need to opt for geometrically and semantically rich digital models that integrate the interrelationships between the different urban components. The literature review by Biljecki *et al.* (2015) identifies at least 29 uses and a hundred applications around 3D urban models, ranging from simple visualizations to more complex analytical systems. The applications either concern one of the components of the urban space (infrastructures and networks, buildings, land and cadastral parcels, etc.) or integrate several components into a single urban application. An example of this is a real estate appraisal, whose current research integrates indoor (BIM) and outdoor (CIM) aspects to study the impact of internal and external parameters on the real estate value of a property (El Yamani *et al.* 2019).

3.3. Semantics of 3D urban models

Several urban models have been developed in the literature to address a wide range of applications. The 3D urban models differ in the amount and granularity of geometric information represented, and in the adopted data structures. Due to the diversity of uses, the semantics of 3D urban models has been the subject of numerous works and publications. Virtual city models, used for urban planning, can be semantically enriched by methods of integration of several data sources (Döllner and Hagedorn 2007). Several approaches have been proposed for the semantization of 3D urban models. In particular, the extraction of semantics from 2D models, maps and MNTs, and the integration of data from several sources for the management of urban spaces.

The need to integrate city semantics into urban models was among the motivations behind the adoption of the CityGML standard (Kolbe 2009). Indeed, CityGML proposes a rich data model, allowing the integration of several urban themes around a geometric core model, in order to perform spatial and thematic analysis tasks. CityGML defines five levels of detail for buildings, ranging from the representation of the footprint (LoD0) to the representation of its detailed architecture containing internal elements, such as rooms and equipment (LoD4). By moving from one level of detail to a higher level, the representation is refined not only by increasing the geometric complexity, but also by the semantic richness. On the other hand, the AEC domain has developed the IFC standard, which proposes a very rich geometric and semantic architecture for the intelligent management of a building, covering all its structural and functional components.

The interconnection between several urban models allows for a better representation of reality and planning according to a global approach (Métral *et al.* 2009). With the integration of CityGML and IFC standards, digital mock-ups are extended to multi-scale applications, taking advantage of the richness of building modeling provided by IFC and the richness of their urban space provided by CityGML within the same application. However, this integration poses interoperability problems that have yet to be resolved.

3.4. From the point cloud to the 3D model

The sources of 3D data are numerous and continue to develop. However, lasergrammetry is well suited to the development of BIM models at the building scale and large urban scale (multi-scale BIM), while considering other sources such as photogrammetry, RGB-D sensors and tacheometric methods, which often remain as complementary methods.

3.4.1. *Point cloud processing chain*

With the evolution of lasergrammetry scanning technologies, point clouds are a potential source for the production of digital models. The point cloud has certain characteristics such as density, presence of noise, occlusions, etc., that must be taken into account in the processing.

The essential steps in the process leading from the cloud to a BIM model can be broken down into four phases: (1) consolidation and/or georeferencing; (2) pre-processing; (3) segmentation/classification; and (4) modeling (Figure 3.1).

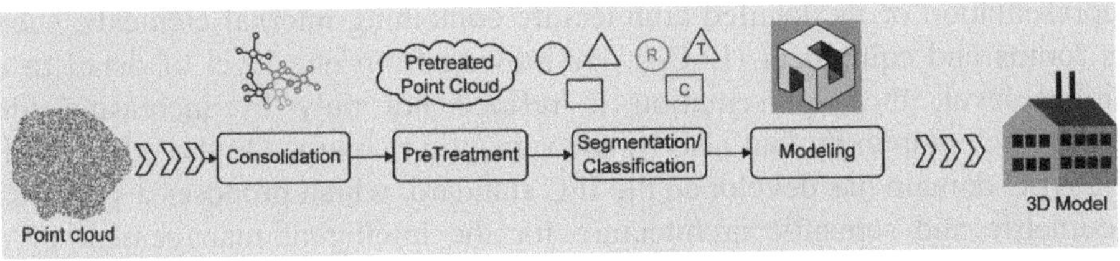

Figure 3.1. *Processing chain from a point cloud to a 3D model. For a color version of this figure, see www.iste.co.uk/hajji/BIM.zip*

3.4.1.1. *Consolidation and georeferencing*

Point cloud registration begins with a consolidation step (also called 3D registration) to assign correspondences between two sets of points. This consolidation is done by identifying homologous points in the various clouds to be consolidated, which can either be targets or homologous geometric entities (planes, spheres, cylinders), or on the basis of specific algorithms such as the DARCES method: Data-Aligned Rigidity-Constrained Exhaustive Search based on RANSAC (RANdom SAmple Consensus) and ICP (Iterative Closest Point) (Landes *et al.* 2011).

To be able to link a point cloud to a national geographic reference frame, indirect georeferencing is required. In this case, topographic techniques such as tacheometry or GNSS surveys are used to link the clouds to the base geographic reference frame. As an alternative to this indirect solution, some scanners allow direct georeferencing on site through a polygonal path scan (Figure 3.2).

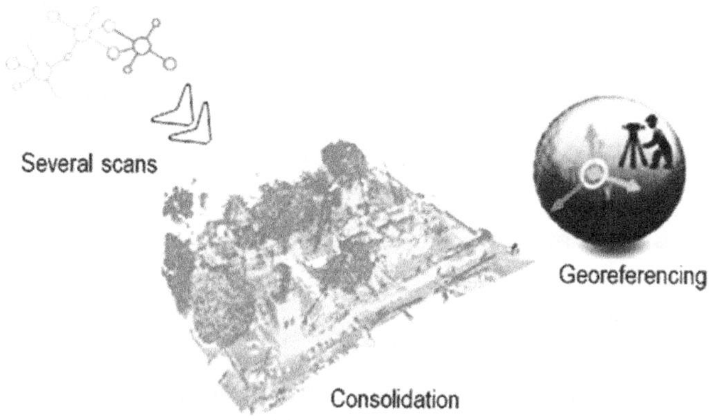

Figure 3.2. *Consolidation and georeferencing of a point cloud. For a color version of this figure, see www.iste.co.uk/hajji/BIM.zip*

3.4.1.2. *Pre-treatment*

Once the various point clouds have been assembled in the same frame and the results of the consolidation have been validated, a point cloud cleaning step is necessary to eliminate undesirable points that add noise to the measurements (Figure 3.3). This step is either done manually by using the tools available in point cloud processing software, or automatically using filters that allow the removal of isolated points based on the distance from their surroundings.

Cleaning can also be done with machine learning-based algorithms such as PointCleanNet (Rakotosaona and Polytechnique 2019). More recently, a part of this operation can be done by acquiring two scans one after the other to minimize the effect of objects moving during the survey (the double scan).

At the end of the pre-processing, the point cloud is consolidated, cleaned and sometimes georeferenced. It can then enter a process of segmentation and modeling to produce a BIM model.

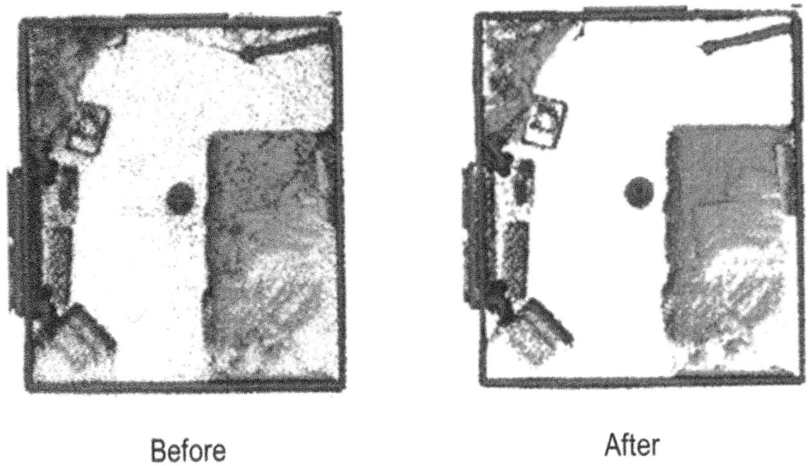

Figure 3.3. *Pre-processing of point clouds by cleaning. For a color version of this figure, see www.iste.co.uk/hajji/BIM.zip*

3.4.1.3. *Segmentation*

The 3D reconstruction of urban models from LiDAR data involves two fundamental steps: (1) segmentation and (2) 3D modeling of the resulting classes.

Segmentation is based on the homogeneous characteristics of the scene data with respect to a certain number of criteria in order to separate them into several regions (Deveau 2006). It involves decomposing a cloud into subsets based on similarity or dissimilarity criteria, (Figure 3.4). In the case of point clouds, the 3D position of points, curvature or flatness are criteria that are considered in the segmentation process (Boulaassal 2010). Other criteria concern geometric, spectral, semantic characteristics, etc.

The process of segmentation is well known in the field of image processing. The segmentation algorithms were originally developed and applied to images. Their extension to the case of point clouds by airborne LiDAR is argued by the fact that airborne LiDAR acquisitions are of the 2.5D type, and their transformation into images does not cause a loss of consistent information that could alter their accuracy. However, in the case of terrestrial LiDAR, the scene is acquired in 3D. Hence, the need to adapt the models and algorithms to take into account the 3D character of the scene.

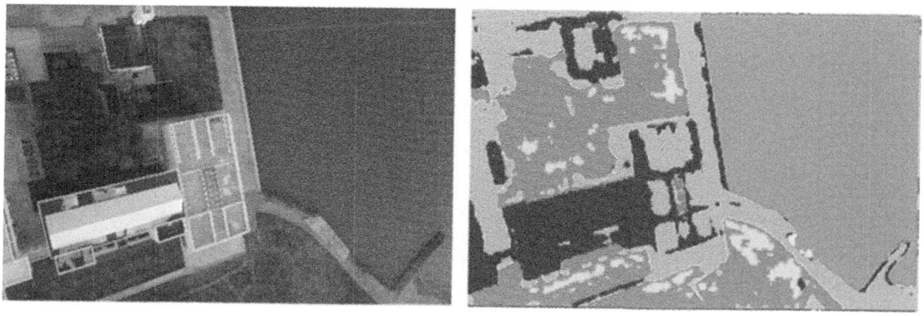

a) b)

Figure 3.4. *The principle of segmentation. For a color version of this figure, see www.iste.co.uk/hajji/BIM.zip*

Segmentation approaches

Many segmentation approaches are reported in the literature with several classifications. A first classification allows the following main families of methods to be distinguished: (1) edge-based segmentation; (2) region growing segmentation; (3) model fitting segmentation; and (4) machine learning segmentation (Figure 3.5). A summary of these approaches was provided by Grilli *et al.* (2017).

Another classification classifies LiDAR data segmentation methods into three main families of approaches: (1) direct approaches based on the raw point cloud; (2) derived approaches based on a cloud-derived product; and (3) hybrid approaches combining several data sources (LiDAR, aerial images, 2D maps, etc.). A state of the art on these different approaches is reported by Ballouch *et al.* (2020).

The segmentation approaches developed in the literature are based on LiDAR data acquired from several platforms and concern both the interior and exterior of the building.

In the case of airborne LiDAR, we cite the example of Tarsha-Kurdi *et al.* (2008), who developed a semi-automated approach for the segmentation of point clouds on an urban scene, where a first distinction between the ground and aboveground is made through the generation of an DSM (Digital Surface Model) (Figure 3.6), and then an extraction of buildings from the ground segment is conducted by adopting a region growth algorithm.

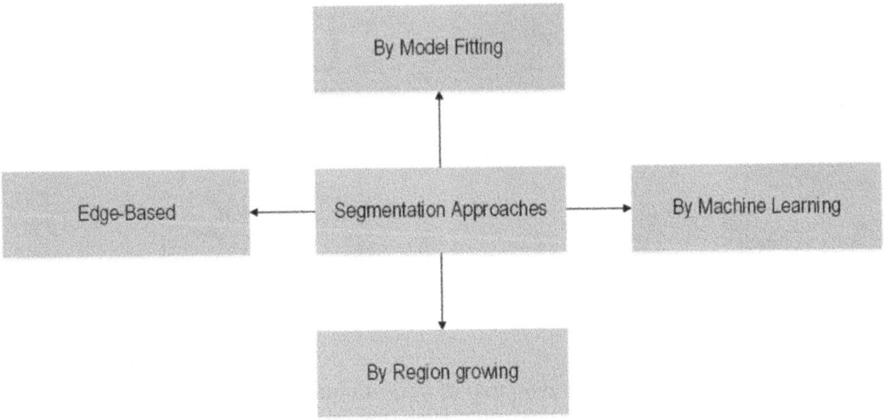

Figure 3.5. *Segmentation approaches*

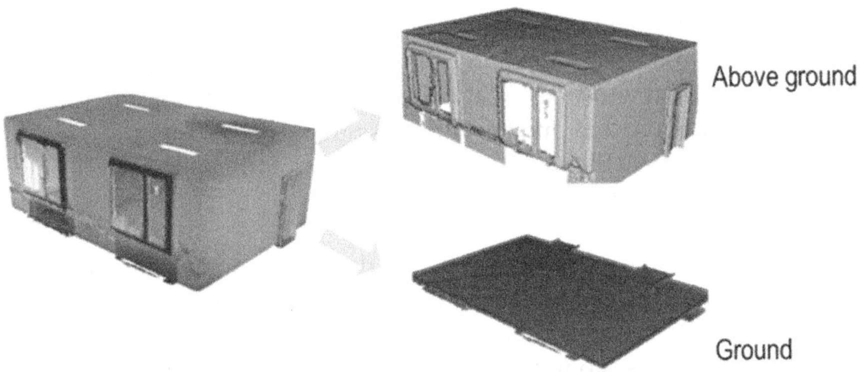

Figure 3.6. *Ground/above ground segmentation. For a color version of this figure, see www.iste.co.uk/hajji/BIM.zip*

In the case of terrestrial LiDAR, Boulaassal (2010) proposes a semi-automatic segmentation of building facades surveyed by a static terrestrial scanner. The planar segments corresponding to the building facades are extracted by the RANSAC algorithm, which has demonstrated its robustness, even in the presence of outliers in the point cloud.

For the case of indoor acquisitions, Anand *et al.* (2011) propose a segmentation approach based on the use of contextual data, such as local visual appearance, shape cues and geometric relationships, to semantically segment point clouds of building interiors.

Advances in segmentation methods are more notable in the image mode than in the point cloud mode, given the time lag in the development of the two acquisition methods. Thus, several studies attempt to exploit image-based approaches by projecting the point cloud onto images. One approach, proposed by Tabkha *et al.* (2019), uses the 360° panoramic image based on cylindrical projections to transform a point cloud into images and apply image-based segmentation processes to them.

Semantic segmentation (classification)

When the cloud segmentation approach is not controlled by prior knowledge, the segmentation results lack semantic information (Xie *et al.* 2019). Once the segmentation process is accomplished, each cluster of points can be semantically augmented by associating a semantic label with it. This is a semantization of the clouds that occurs after the segmentation process. This process is called semantic segmentation. Semantic segmentation is known as classification in the field of photogrammetry and remote sensing.

In computer vision, semantic segmentation is based on subdividing a point cloud into parts with a semantic assignment by associating semantic labels with each class (Lawin *et al.* 2017) (Figure 3.7). Semantic segmentation of 3D point clouds is a major challenge for understanding and extracting knowledge from urban scenes. This is one of the major challenges in the field of computer vision (Ballouch *et al.* 2020).

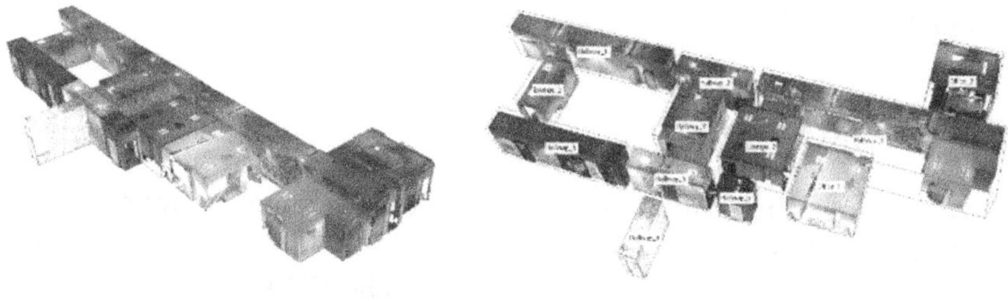

Figure 3.7. *Semantic segmentation/classification of a scene. For a color version of this figure, see www.iste.co.uk/hajji/BIM.zip*

Semantic segmentation approaches are distinguished into: (1) supervised approaches; (2) unsupervised approaches; and (3) interactive approaches. Supervised approaches use data sets or training models for semantic classification of the cloud. These approaches require massive training data to train the classification model. Unsupervised approaches are automatic methods based on user parameterization of the classification algorithm and can lead to imprecise results, while interactive approaches require user intervention to guide the segmentation process (Grilli *et al.* 2017).

Today, deep learning methods and Convolutional Neural Networks (CNN) allow us to solve many problems related to the thematic exploitation of massive point clouds (image classification and analysis, real-time object detection on videos, semantic segmentation, Natural Language Processing (NLP), audio recognition, etc.). The term Deep Learning (DL) refers to the training of neural networks. It is one of the most important forms of artificial intelligence derived from machine learning, whose basic architecture is inspired by the biological processes of connections between neurons in the brain, assembled together.

The advantage of a CNN is the preservation of the spatial information of the visual data, and also the considerable decrease of the weight and bias parameters, which allows a faster execution and more accurate predictions. The algorithm mainly relies on the detection of features (boundaries, shapes, texture, etc.) in the images through filters based on the convolution operation.

Several semantic segmentation approaches based on DL have been developed in the literature. A state of the art of these approaches is reported in Ballouch *et al.* (2020).

3.4.2. *Geometric modeling*

To overcome the limitations of traditional geometric representations, BIM offers an integrated data model that models both the elements of the building and the relationships between them. The BIM model centralizes all information about the building and/or other urban objects. This information includes geometric aspects concerning the elements of the model, as well as semantic aspects which can relate to different themes.

The BIM model adopts a parametric object-oriented modeling where each object of the model has a geometric definition and encapsulates parameters and functions. The scan-to-BIM process allows us to produce the model from a laser scan through a 3D reconstruction approach. The development of automatic approaches for the scan-to-BIM process is a very active research area.

3.4.2.1. *Geometric modeling methods*

3D geometric modeling methods are distinguished by the type of geometric primitives used and the relations of the combination adopted to associate them, in order to build a 3D shape. Their choice depends on the intended application: some models are efficient for 3D visualization, others perform well in spatial analysis and 3D simulation applications.

Many 3D geometric models have been proposed in the literature. A first classification of 3D modeling methods by Li (1994) distinguishes the methods into two main categories: (1) surface methods when the object is reconstructed based on surface primitives and (2) volume methods when the object is reconstructed based on solids. Another category classifies the models into: (1) models based on geometric primitives; (2) mesh models; and (3) hybrid models.

Surface methods

The main surface methods are: 2D grid mode, triangulations, NURBS (Non-Uniform Rational B-Splines) methods, B-REP (Boundary Representation) method, etc.

NURBS and B-Rep methods are well suited for modeling buildings and other urban objects. NURBS modeling is based on curves that are defined by control points and constraints imposed on a curve through a precise mathematical definition. The B-Rep method is a boundary-based modeling that describes a 3D shape by the surfaces composing its boundaries: nodes, edges and knots. This method integrates an implicit topological representation.

Volume methods

In this category of methods, we find the CSG (Constructive Solid Geometry) method, the SPR (Spatial Partitioning Representation) method, Primitive Instancing (PI), etc.

The CSG model is based on the combination of geometric primitives for the construction of complex objects through geometric transformations or Boolean operations (union, intersection, difference, etc.). This allows the construction of a good number of urban objects with complex geometry. The SPR method decomposes an object into a set of juxtaposed elementary solids that do not intersect and that can vary in terms of types, sizes, positions, parameters and orientations. In this category, we find the voxel model, which consists of splitting a solid into elementary cubes, and the Octree model, which is an improvement of the "Voxel" model. Octrees allow us to represent objects by a hierarchical succession of cubes of variable sizes. The Octree model allows a hierarchical representation of the object which adapts to a display at different resolutions.

The "Primitive Instancing" method uses complex parameterizable primitives to represent a 3D scene, which are usually stored in an object library.

Mesh modeling

The 3D mesh is one of the most popular representations of 3D shapes. A 3D mesh structure describes the geometry and appearance of the objects it represents with an associated texture. It consists of a set of polygons called faces, described by a diagram showing the location and connectivity of their vertices. The objects can be easily recognized by the human eye, but cannot be distinguished by computer systems (Willenborg *et al.* 2018). 3D meshes can also be presented as graphically structured data where the nodes of the graph correspond to the vertices of the mesh and the edges represent the connectivity between these vertices. This mode is usually combined with the point cloud mode where the points of the cloud form the vertices of the triangles which are calculated mathematically, in most cases, by Delaunay triangulation. Thus, the mesh, bringing no additional information, remains more easily interpretable than a point cloud with a good visual rendering and a realistic visualization quality. The object format (OBJ) is one of the most popular 3D formats for storing and exchanging 3D mesh models that defines a geometry suitable for 3D modeling and computer graphics (Willenborg *et al.* 2018).

Parametric modeling

BIM modeling is based on the definition of object classes containing instances whose appearance changes based on parameters associated with the class (common parameters) or object (object-specific parameters), and also the relationships they undertake with other objects. This is the principle of parametric modeling, where representation is made based on parameters and rules that determine the geometric and non-geometric properties of objects (Eastman *et al.* 2011).

Parametric modeling consists of associating a set of parameters (qualitative and/or quantitative) with an object, that can easily be modified (Figure 3.8). This modeling defines the relationships between the components of an object, which can, for example, be represented by mathematical equations. The set of rules assigned to the objects allows them to be automatically updated following a change of context.

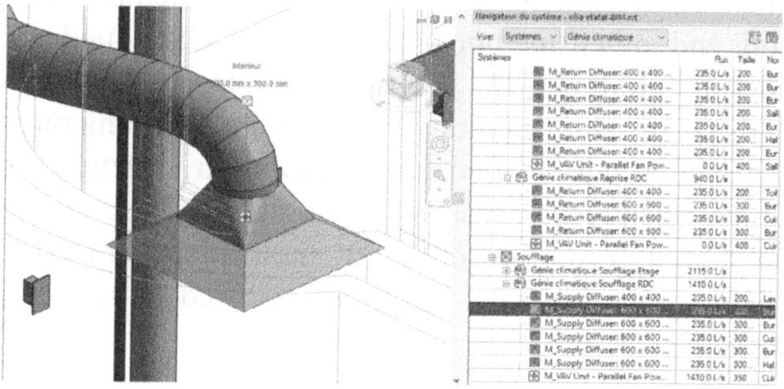

Figure 3.8. *Parametric modeling of a BIM object. For a color version of this figure, see www.iste.co.uk/hajji/BIM.zip*

With BIM tools, it is possible to create libraries of intelligent objects that can store and use information about the project. For example, a door, placed in a BIM model, is a simple assembly of lines and surfaces, but represents an object connected to a wall and contains information about materials, costs, suppliers, etc.

3.4.2.2. *Selection criteria*

The choice of a 3D geometric modeling must allow the representation of a large number of urban objects, a multi-level representation, a storage performance and a 3D spatial analysis (Koussa 2011). In addition, the adopted model must be able to be translated to the schema of a 3D standard, such as CityGML and IFC, to ensure interoperability of exchanges.

In the case of buildings, surface representations such as B-Rep and NURBS are the best known. The B-Rep method, by its intuitive character, allows us to cover a very large domain since most solids can be represented by surfaces. It is well adapted for multi-scale use and has a good topological potential. In terms of interoperability, it is compatible with the geometric primitives (Point, LineString, Polygon, MultiLineString, Multi Polygon, etc.) defined by the OGC specifications. In addition, the B-Rep method can allow 3D functions such as volume calculation in the case where the faces are oriented (Ramos 2003). However, the method is not very compact since the description of some objects requires a large number of surfaces to represent them.

In the urban environment, the use of hybrid methods combining the B-Rep method and other methods such as the CSG method allows the representation of buildings to be extended to model more complex geometries. Most of the methods use variants of hierarchical space partitioning structures, also called multi-resolution structures, such as k-d trees, octrees or quadtrees. Massive point clouds need to make use of suitable structures in order to optimize their management. One of the examples of these structures is the Modifiable Nested Octree (MNO).

3.5. 3D reconstruction of the BIM model

In this section, we focus on the 3D reconstruction of objects for a BIM model. We focus on recent developments in this area which are oriented towards the exploitation of point clouds from lasergrammetry or low-cost sensors for indoor reconstruction of buildings. We also refer to the reconstruction of multi-scale BIM models from large-scale acquisition techniques, such as drones or airborne LIDARs.

3.5.1. *Scan-to-BIM process*

3D scanning technologies have brought a very significant evolution to the methods of modeling geometric information and integrating it into a BIM model. The process of obtaining a BIM model from a point cloud (scan) is called "scan-to-BIM". The point cloud resulting from a scan is the basis for the reconstruction of an "As-Built" BIM.

Currently, terrestrial lasergrammetry is a potential technique for the acquisition of large point clouds on buildings and other urban objects. The technique generates millions of points acquired in a short time with high resolution and geometric accuracy. Other mobile scanning techniques are also used in indoor (SLAM) or outdoor (MMS) environments. These optimize the acquisition time, but their quality is affected by the movement of the platform and thus present limitations in front of applications requiring high geometric accuracy (Thomson *et al.* 2013).

Currently, scan-to-BIM is a manual process conducted by "expert modelers" who rely on the point cloud to model the objects of the model (Bassier *et al.* 2020). The most common approach is to import the point cloud into a BIM modeling software (Revit, Bentley, Graphisoft, ArchiCAD, etc.) and create the model using the point cloud as a reference. However, this method is laborious, a source of errors and unsuitable for large projects generating large amounts of data, hence the interest of the research community in developing automatic approaches for the scan-to-BIM process.

The scan-to-BIM process is performed through several steps: segmentation, classification, clustering and finally, the creation of the model. The objects resulting from the segmentation/classification can then be converted into BIM objects by associating semantic descriptions with them (Figure 3.9).

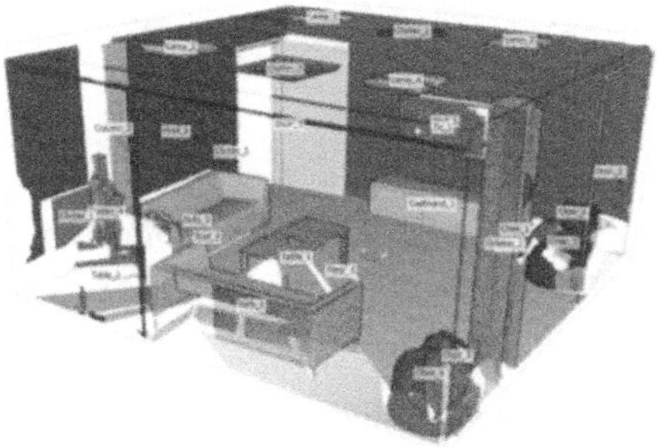

Figure 3.9. *Classification of a point cloud in an indoor environment. For a color version of this figure, see www.iste.co.uk/hajji/BIM.zip*

The object modeling in a BIM model requires a preliminary step of segmentation and reconstruction of elements such as walls, openings (doors and windows), rooms, etc. This segmentation is practiced according to several approaches. The works in this context are interested in either the modeling of spaces or that of elements composing the building, such as walls (Macher *et al.* 2017).

3D space reconstruction

A simple but critical example of the difference between a geometric data model and a BIM model is the representation of the concept of space (Figure 3.10). This concept is not necessarily represented by a geometric definition stored in the model, but "space" in a BIM model has a semantic definition that relates to a domain and can be extracted by a simulation tool. This concept is used for certain applications such as energy simulations. Indeed, the open spaces that are required for the calculation of energy losses can be easily obtained from a BIM model.

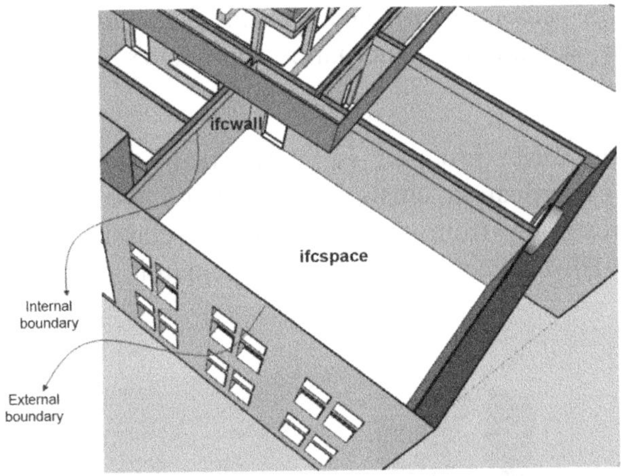

Figure 3.10. *The notion of space in IFC. For a color version of this figure, see www.iste.co.uk/hajji/BIM.zip*

Several approaches are developed to segment subspaces in a building. Some authors use histograms for a vertical segmentation of the point cloud (Macher *et al.* 2014). Díaz-Vilariño *et al.* (2015) use a histogram of point distribution along the Z-axis to identify the different floors of a building. Ochmann *et al.* (2014) propose a segmentation into rooms and a location of openings by determining the probability that each point has of belonging to a room. Mura *et al.* (2014) propose a modeling of rooms as polyhedra by starting with the location of walls and their projection in the X–Y plane, and then a decomposition into cells is performed by the intersection of lines.

Reconstruction of elements

Among the approaches that have been developed for wall segmentation, we mention the one proposed by Hong *et al.* (2015) that segments walls based on 2D raster image creation for the roofs and floors of the building. Wang *et al.* (2015) use machine learning methods and neural networks for wall segmentation. Thomson and Boehm (2015) propose an approach based on horizontal plane segmentation for floor/ceiling separation, followed by vertical plane segmentation for wall identification.

In the literature, several authors have developed semi-automatic or automatic approaches for BIM model reconstruction, including Xiong *et al.* (2013), Ochmann (2019) and Bassier *et al.* (2020). A review of the literature on scan-to-BIM automation techniques is reported by Tang *et al.* (2010) and Pătrăucean *et al.* (2015).

Recently, other authors such as Li *et al.* (2020) have addressed the use of low-cost sensors such as RGB-D cameras for automatic as-built BIM generation. Bassier *et al.* (2019) developed an unsupervised method for wall reconstruction from laser scans and the generation of BIM objects within Revit after the segmentation of the cloud by machine learning algorithms. However, very few works have exploited the radiometric values of the clouds. Notably, Zhan *et al.* (2009) propose a method of contour extraction based on the colorimetric differences in a point cloud colored from RGB information.

Research for BIM reconstruction approaches from point clouds is still relevant. However, the quality assessment of the generated models does not require the same order of interest by the researches. Only a few researchers have addressed the issue of model quality evaluation (Previtali *et al.* 2014; Tran *et al.* 2019). Recently, Bassier *et al.* (2020) proposed a comparison of the reconstructed objects to the initial point cloud in terms of LoA (Level of Accuracy) and LoD (Level of Development).

3.5.2. *Scan versus BIM process*

In contrast to the scan-to-BIM process, which transforms discrete information from a point cloud into modeled information in the form of BIM objects, scans can also be used to control a BIM model. Among the uses of scans versus BIM, we cite the superposition between periodic scans and the BIM model to calculate the progress of operations in relation to an existing schedule (BIM 4D) (Son *et al.* 2017), and to verify possible discrepancies through a geometric quality control, allowing us to control the progress of operations and take corrective measures in case of problems. Some software offer functionalities and plugins allowing a geometric registration of scans and a dimensional analysis that can be performed for a gap analysis between a scan, and a reference surface for the inspection of building facades, for example, or between several scans for a spatio-temporal analysis of movements of a structure.

3.6. Conclusion

Today, the production of 3D digital models, whether on a building scale (BIM model) or on a large scale (multi-scale BIM), is a major challenge in order to respond to several urban management and smart city development issues.

The point clouds produced from different sources offer an important database for the elaboration of digital models with high geometric and semantic accuracy. The scan-to-BIM processes still need to be developed to ensure sufficient automation in the production of models. Despite the developments, the semantization of point clouds is still a very active field of research that takes advantage of the development of DL algorithms.

3.7. References

Abdul-Rahman, A. and Pilouk, M. (2007). *Spatial Data Modelling for 3D GIS*. Springer Science & Business Media, Berlin.

Anand, A., Koppula, H.S., Joachims, T., Saxena, A. (2012). Contextually guided semantic labeling and search for three-dimensional point clouds. *The International Journal of Robotics Research*, 32(1), 19-34.

Ballouch, Z., Hajji, R., Ettarid, M. (2020). The contribution of deep learning to the semantic segmentation of 3D point-clouds in urban areas. *2020 IEEE International Conference of Moroccan Geomatics*, IEEE, Morgeo, 1–6.

Bassier, M., Mattheuwsen, L., Vergauwen, M. (2019). Bim reconstruction: Automated procedural modeling from point cloud data. *The International Archives of Photogrammetry, Remote Sensing and Spatial Information Sciences*, 42, 53–60.

Bassier, M., Yousefzadeh, M., Vergauwen, M. (2020). Comparison of 2D and 3D wall reconstruction algorithms from point cloud data for as-built BIM. *Journal of Information Technology in Construction*, ITcon, 25(11), 173–192.

Biljecki, F., Stoter, J., Ledoux, H., Zlatanova, S., Çöltekin, A. (2015). Applications of 3D city models: State of the art review. *ISPRS International Journal of Geo-Information*, 4(4), 2842–2889.

Boulaassal, H. (2010). Segmentation et modélisation géométriques de façades de bâtiments à partir de relevés laser terrestres. Doctoral Dissertation, University of Strasbourg.

Deveau, M. (2006). Utilisation conjointe de données image et laser pour la segmentation et la modélisation 3D. Doctoral Dissertation, Paris Descartes University.

Díaz-Vilariño, L., Conde, B., Lagüela, S., Lorenzo, H. (2015). Automatic detection and segmentation of columns in as-built buildings from point clouds. *Remote Sensing*, 7(11), 15651–15667.

Döllner, J. and Hagedorn, B. (2007). Integrating urban GIS, CAD, and BIM data by service-based virtual 3D city-models. In *26th Urban and Regional Data Management: UDMS 2007 Annual*, Rumor, M., Coors, V., Fendel, E.M., Zlatanova, S. (eds). Taylor & Francis Ltd., Stuttgart.

Eastman, C.M., Eastman, C., Teicholz, P., Sacks, R., Liston, K. (2011). *BIM Handbook: A Guide to Building Information Modeling for Owners, Managers, Designers, Engineers and Contractors*. John Wiley & Sons Inc, Hoboken, NJ.

El Yamani, S., Ettarid, M., Hajji, R. (2019). BIM potential for an enhanced real estate valuation approach based on the hedonic method [Online]. Available at: http://hdl.handle.net/2268/250013.

Grilli, E., Menna, F., Remondino, F. (2017). A review of point clouds segmentation and classification algorithms. *Proceedings of the International Archives of the Photogrammetry, Remote Sensing and Spatial Information Sciences – ISPRS Archives*, 42, 339–344.

Hong, S., Jung, J., Kim, S., Cho, H., Lee, J., Heo, J. (2015). Semi-automated approach to indoor mapping for 3D as-built building information modeling. *Computers, Environment and Urban Systems*, 51, 34–46.

Kolbe, T.H. (2009). Representing and exchanging 3D city models with CityGML. *3D Geo-Information Sciences*. Springer, Berlin, Heidelberg.

Koussa, C. (2011). Implantation d'un système d'information géographique 3D sur Internet pour la gestion des modèles urbains. Doctoral Dissertation, University of Strasbourg.

Landes, T., Grussenmeyer, P., Boulaassal, H. (2011). Les principes fondamentaux de la lasergrammétrie terrestre : acquisition, traitement des données et applications, article XYZ (partie 2/2), 25–38.

Lawin, J.-L., Danelljan, M., Tosteberg, P., Bhat, G., Khan, F.S., Felsberg, M. (2017). Deep projective 3D semantic segmentation. In *Computer Analysis of Images and Patterns: 17th International Conference, CAIP 2017*, Felsberg, M., Heyden, A., Krüger, N. (eds). Springer, Cham.

Li, R. (1994). Data structures and application issues in 3-D geographic information systems. *Geomatica*, 48(3), 209–224.

Li, Y., Li, W., Tang, S., Darwish, W., Hu, Y., Chen, W. (2020). Automatic indoor as-built building information models generation by using low-cost RGB-D sensors. *Sensors*, 20(1), 293.

Macher, H., Landes, T., Grussenmeyer, P., Alby, E. (2014). Boundaries extraction from segmented point clouds as input for historical building information modelling. *International Journal of Heritage in the Digital Era*, 3(4), 669–682.

Macher, H., Landes, T., Grussenmeyer, P. (2017). From point clouds to building information models: 3D semi-automatic reconstruction of indoors of existing buildings. *Applied Sciences*, 7(10), 1030.

Métral, C., Falquet, G., Cutting-Decelle, A. (2009). Towards semantically enriched 3D city models: An ontology-based approach. *International Archives of Photogrammetry, Remote Sensing and Spatial Information Sciences*, 38, (Parts 3–4/C3) (on CD–ROM).

Mura, C., Mattausch, O., Villanueva, A.J., Gobbetti, E., Pajarola, R. (2014). Automatic room detection and reconstruction in cluttered indoor environments with complex room layouts. *Computers & Graphics*, 44, 20–32.

Ochmann, S.K. (2019). Automatic reconstruction of parametric, volumetric building models from 3D point clouds. Doctoral Dissertation, Universitäts-und Landesbibliothek Bonn.

Ochmann, S.K., Vock, R., Wessel, R., Tamke, M., Klein, R. (2014). Automatic generation of structural building descriptions from 3D point cloud scans. *2014 International Conference on Computer Graphics Theory and Applications (GRAPP)*, IEEE, 1–8.

Pătrăucean, V., Armeni, I., Nahangi, M., Yeung, J., Brilakis, I., Haas, C. (2015). State of research in automatic as-built modelling. *Advanced Engineering Informatics*, 29(2), 162–171.

Previtali, M., Scaioni, M., Barazzetti, L., Brumana, R. (2014). A flexible methodology for outdoor/indoor building reconstruction from occluded point clouds. *ISPRS Annals of the Photogrammetry, Remote Sensing and Spatial Information Sciences*, 2(3), 119.

Rakotosaona, M.J., La Barbera, V., Guerrero, P., Mitra, N.J., Ovsjanikov, M. (2020). PointCleanNet: Learning to denoise and remove outliers from dense point clouds. *Computer Graphics Forum*, 39(1), 185–203.

Ramos, F. (2003). Modélisation et validation d'un système d'information géographique 3D opérationnel. Doctoral Dissertation, Université de Marne-la-Vallée.

Son, H., Kim, C., Kwon Cho, Y. (2017). Automated schedule updates using as-built data and a 4D building information model. *Journal of Management in Engineering*, 33(4), 04017012.

Tabkha, A., Hajji, R., Billen, R., Poux, F. (2019). Semantic enrichment of point cloud by automatic extraction and enhancement of 360° panoramas. *International Archives of the Photogrammetry, Remote Sensing and Spatial Information Sciences*, XLII-2/W17, 355–362 [Online]. Available at: https://doi.org/10.5194/isprs-archives-XLII-2-W17-355-2019.

Tang, P., Huber, D., Akinci, B., Lipman, R., Lytle, A. (2010). Automatic reconstruction of as-built building information models from laser-scanned point clouds: A review of related techniques. *Automation in Construction*, 19(7), 829–843.

Tarsha-Kurdi, F., Landes, T., Grussenmeyer, P. (2008). Extended RANSAC algorithm for automatic detection of building roof planes from LIDAR data. *The Photogrammetric Journal of Finland*, 21(1), 97–109.

Thomson, C. and Boehm, J. (2015). Automatic geometry generation from point clouds for BIM. *Remote Sensing*, 7(9), 11753–11775.

Thomson, C., Apostolopoulos, G., Backes, D., Boehm, J. (2013). Mobile laser scanning for indoor modelling. *ISPRS Annals of the Photogrammetry, Remote Sensing and Spatial Information Sciences*, 5(W2), 66.

Tran, H., Khoshelham, K., Kealy, A. (2019). Geometric comparison and quality evaluation of 3D models of indoor environments. *ISPRS Journal of Photogrammetry and Remote Sensing*, 149, 29–39.

Wang, X., Fouhey, D., Gupta, A. (2015). Designing deep networks for surface normal estimation. *Proceedings of the IEEE Conference on Computer Vision and Pattern Recognition*, 539–547.

Willenborg, B., Pültz, M., Kolbe, T.H. (2018). Integration of semantic 3D city models and 3D mesh models for accuracy improvements of solar potential analyses. *International Archives of the Photogrammetry, Remote Sensing & Spatial Information Sciences*, Volume XLII-4/W10, 1–2 October, Delft.

Xie, Y., Tian, J., Zhu, X.X. (2020). Linking points with labels in 3D: A review of point cloud semantic segmentation. *IEEE Geoscience and Remote Sensing Magazine*, 8(4), 38–59.

Xiong, X., Adan, A., Akinci, B., Huber, D. (2013). Automatic creation of semantically rich 3D building models from laser scanner data. *Automation in Construction*, 31, 325–337.

Zhan, Q., Liang, Y., Xiao, Y. (2009). Color-based segmentation of point clouds. *Laser Scanning*, 38(3), 155–161.

4

Open BIM Standards

In the context of the implementation of an interoperable multi-scale BIM, it is essential to use standardized exchange formats, most often resulting from standardization organizations (OGC, ISO and buildingSMART). These standards allow the implementation of an autonomous model, independent of the tools and software used.

4.1. Standardization bodies

4.1.1. *Open Geospatial Consortium*

The Open Geospatial Consortium (OGC) is an international organization whose main objective is to develop interoperable standards for geospatial data. It has more than 500 members (from industry, academia, government agencies, universities, laboratories, etc.). Spatial information and processing includes Geographic Information Systems (GIS), remote sensing, surveying and mapping, navigation, access to spatial databases, sensor networks and other geospatial technologies and information sources for terrestrial BIM implementation. In the OGC consensus process, more than 360 governments and private and cooperative organizations develop, test, document, validate and approve interface and interoperability issues. The OGC baseline of adopted standards includes these implementation specifications.

OGC has a Class A liaison with ISO TC 211, whereby some OGC specifications become ISO standards (Figure 4.1). These two organizations

---

Chapter written by Hassane JARAR OULIDI and Rafika HAJJI.

have developed an agreement to cooperate and harmonize their work. This is the case for the Geography Markup Language (GML) specifications that are the basis of ISO standards 19136 and 19128.

The OGC is responsible for, among others, the following standards:

– Geography Markup Language (GML) and CityGML;

– GeoScience Markup Language (GeoSciML);

– Geospatial User Feedback (GUF) and so on.

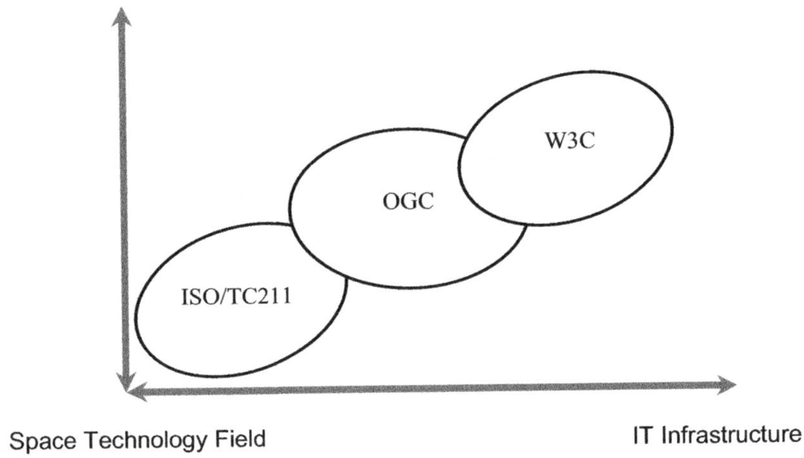

Figure 4.1. *Relationship between OGC and other geospatial information standards organizations*

4.1.2. *ISO TC211*

TC 211 (Geographic Information/Geomatics) was created in 1994 and is attached to the International Organization for Standardization (ISO), of which ANSI is the official representative in the USA. TC 211 has established a series of ISO standards that are specific to geographic information. In view of the large number of industries in the field of geographic information systems and the growing interest in opening up and exchanging data between different systems, an industry consortium was created in 1994, the Open Geospatial Consortium (OGC) which currently includes 477 companies, government agencies and universities. Beginning in 1998, the OGC and TC 211 signed a cooperative agreement, and as part of this agreement, the OGC adopted a significant number of TC 211 standards. These standards include rules and techniques necessary for modeling, acquiring, archiving and visualizing geographic information.

4.1.3. *buildingSMART*

buildingSMART is an international consortium of partners operating in the AEC sector. It is committed to creating Open BIM digital methods to ensure interoperability. buildingSMART standards help construction project stakeholders work more efficiently and collaboratively throughout the lifecycle of the project.

The buildingSMART organization has developed five standards for working in an Open BIM environment. These standards are as follows: Information Dissemination Manual (IDM), Model View Definition (MVD), Digital Dictionary Registry (DDR), BIM Collaboration Format (BCF) and IFC Data Storage and Structuring Model.

4.2. Data models for multi-scale BIM

4.2.1. *Open BIM*

Industry Foundation Classes (IFCs)

IFCs (Industry Foundation Classes) represent an open specification for Building Information Modeling (BIM). It is data that is exchanged and shared between the different stakeholders of an AEC project. IFCs are the main international standard for Open BIM. IFC specifies a data schema and an exchange file format structure. The data schema is defined in:

– The EXPRESS data specification language, defined in ISO 10303-11.

– The XML Schema Definition Language (XSD), defined in the W3C XML Schema (the EXPRESS schema definition represents the source, while the XML schema definition is generated from the EXPRESS schema according to the rules defined in ISO 10303-28).

The IFC data schema architecture defines four conceptual layers, and each individual schema is assigned to exactly one conceptual layer:

– Resource layer: the lowest layer includes all individual schemas containing resource definitions. These definitions do not include a globally unique identifier and should not be used independently of a definition declared at a higher layer.

– Core layer: the next layer includes the "kernel" schema and the extension schemas, containing the most general entity definitions. All entities defined at the core layer level or above carry a globally unique identifier and possibly owner and history information.

– Interoperability layer: the next layer includes schemas containing entity definitions that are specific to a general product, process or resource specialization used across disciplines. These definitions are typically used for cross-domain exchange and sharing of construction information.

– Domain layer: the highest layer includes schemas containing definitions of entities that are discipline-specific specializations of products, processes or resources. These definitions are typically used for intra-domain exchange and information sharing (see https://standards.buildingsmart.org/ for more information).

The latest official version of IFC is IFC4 ADD2 TC1 (version 4.0.2.1) certified to ISO 16739-1: 2018. However, other versions are being developed to expand the scope of IFC to include, in addition to building-related data definitions, infrastructure-related data throughout its lifecycle.

A project called "IFC Infra Overall Architecture" has been launched with the objective of providing a common basis for future IFC extension projects in the infrastructure domain. The project was conducted in close collaboration between buildingSMART International and OGC in order to realize a harmonized conceptual model as a common basis for IFC-Infra and InfraGML.

The IFC-Infra global architecture project has proposed spatial structure concepts to support infrastructure.

The project was released in 2017. In 2018, IFC4.1 was released and is intended to provide a foundation for the various infrastructure domain extensions being developed. Extensions to the IFC4 schema included the following:

– Description of the alignment as a combination of horizontal and vertical alignment.

– Linear placement according to ISO 19148.

– IfcSectionedSolidHorizontal as a new geometry representation that is particularly useful for describing infrastructure facilities.

In 2019, the primary focus of IFC4.2 was to extend the IFC schema to include the description of bridge constructions. Extensions to the IFC4.1 schema included the following:

– The spatial structure was extended by IfcFacility and IfcFacilityPart as a basis for describing the spatial distribution structure of the infrastructure.

– IfcBridge and IfcBridgePart have been added to represent bridges and bridge parts. The elements of bridge have been integrated into a number of predefined building element types.

– The representations IfcBearing, IfcDeepFoundation, IfcVibrationDamper and IfcTendonConduit have been added to represent the bridge elements.

– IfcRelPositions has been added to better support positioning along the alignment.

In 2020, IFC4.3 RC1 was released, which extends the IFC schema to cover the description of infrastructure constructions in the domains of railroads, roads, ports and waterways, including elements common to these domains (buildingSMART 2020).

Information Delivery Manual (IDM)

The standard IDM provides a conceptual framework to effectively improve the quality of information communication between stakeholders by modeling the processes and information transfer flows during the lifecycle of a construction project. A process must imperatively correspond to a use case following the definition of an objective, justifying the relevance of the process.

The IDM has four main components: processes, requirements/references, tasks and deliverables.

A process includes the workflow of tasks and activities carried out for a particular purpose. It aims to give a clear view of the activities and tasks in terms of the involved actors and their roles. A requirement represents the terms of reference that must be considered in order to meet a specific requirement at any stage of the project lifecycle. These data sets should generally be created in accordance with the IFC data structure. However, it is also possible to adopt an IDM-based approach for other open standards such as CityGML/LandInfra.

Model View Definition (MVD)

The MVD represents a subset of the IFC schema, which supports one or more exchange requirements defined in an IDM by facilitating data exchange and avoiding unnecessary or redundant data sharing, according to standardized protocols. MVDs provide users with the ability to exchange Open BIM data in business processes associated with real projects. Therefore, it is important to ensure that the exchange requirements defined from the end user's perspective are achievable in the software environment.

BIM Collaboration Format (BCF)

The BCF represents an XML file for storing information about coordination processes outside of Open BIM models. In other words, the IFC standard allows for data about a construction project, while the BCF provides a protocol for managing communications around Open BIM models. BCF files typically provide details about each task, which is called a rubric in the BCF context. A BCF-based topic allows for the following: task status, recipient and task type. The BCF can be considered as a simplified approach to coding and coordinating comments on a BIM model, which would then facilitate communication between the AEC project stakeholders and improve the level of collaboration.

4.2.2. *CityGML*

CityGML is an OGC standard that supports 3D virtual models of cities and territories (vegetation, buildings, hydrography, etc.) based on the open GML3 model expressed in XML format. The latest version of CityGML is composed of 14 modules:

– CityGML Core: this defines the basic components of the CityGML model.

– Appearance: this provides the means to model the appearances of CityGML elements.

– Building: this allows the representation of thematic and spatial aspects of buildings, parts of buildings, building installations and interior structures of buildings in five levels of detail (LOD 0–4).

– CityFurniture: this is used to represent urban furniture objects in cities (e.g. street lights and traffic signs).

– CityObjectGroup: this provides a grouping concept for CityGML.

– Generics: this provides generic extensions to the CityGML data model that can be used to model and exchange additional attributes and elements not covered by the predefined CityGML thematic classes.

– LandUse: this allows the representation of areas of the Earth's surface that are dedicated to a specific land use.

– Relief: this allows the representation of the terrain in a city model.

– Transportation: this is used to represent the transportation elements in a city, such as roads, railroads or squares.

– Vegetation: this provides thematic classes to represent vegetation objects.

– WaterBody: this represents the thematic aspects and 3D geometry of rivers, canals, lakes and ponds. However, it does not currently integrate any hydrological or dynamic aspects.

– TexturedSurface: this allows us to assign visual appearance properties (color, brightness and transparency) and textures to 3D surfaces. Due to the limitations inherent in its modeling approach, this module has been marked as obsolete and should be removed in future versions of CityGML.

– Tunnel: this facilitates the representation of thematic and spatial aspects of tunnels, tunnel parts, tunnel facilities and tunnel interior structures in four levels of detail (LOD 1–4).

– Bridge: this allows the representation of thematic and spatial aspects of bridges, bridge parts, bridge facilities and interior bridge structures in four levels of detail (LOD 1–4) (Gröger *et al.* 2012).

The last two modules (Tunnel and Bridge) are new in the latest version. Version no. 1.0 included all of the above modules except for these two modules.

This basic structure of the CityGML standard provides a 3D model of all important objects in the city above ground. However, underground infrastructures are still not supported.

Nevertheless, CityGML provides a mechanism that specifies additions to the CityGML data model: Application Domain Extensions (ADE). These additions include the introduction of new properties into existing CityGML classes, such as the number of inhabitants of a building or the definition of new object types. The difference between ADEs and generic objects and attributes is that an ADE must be defined in an additional XML schema definition file. This file must explicitly import the XML schema definition of the extended CityGML modules (Gröger *et al.* 2012).

Many examples of ADEs have been developed. The one that has defined new objects to support underground infrastructure is the CityGML Utility Network ADE.

CityGML Utility Network ADE supports 3D topographic, 3D topological and functional modeling. It supports complex objects and network hierarchies and also includes features, materials and functional aspects of elements. It also provides a homogenized and integrated view of multi-networks while being independent of the network type.

4.2.3. *LandInfra/InfraGML*

LandInfra is an OGC standard that presents and models land and civil engineering infrastructure. Its conceptual model domains include facilities, projects, alignment, road, rail, surveys, terrain features, land division and wet infrastructure (storm drainage, wastewater and water distribution systems). The initial version of this standard includes all of these areas, with the exception of wet infrastructure, which is expected to be published as a future extension.

InfraGML is published as a standard with several parts, see Figure 4.2.

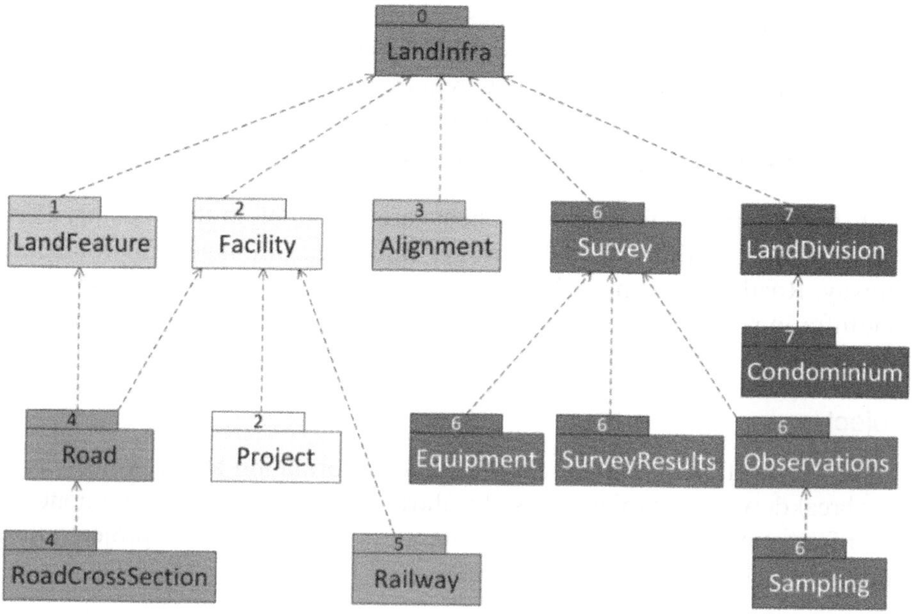

Figure 4.2. *LandInfra classes grouped into InfraGML parts (OGC 2016). For a color version of this figure, see www.iste.co.uk/hajji/BIM.zip*

Part 0: Core

This class contains soil and infrastructure data that may contain information about facilities, land features, land division, documents, survey markers, surveys and feature associations.

Part 1: LandFeature

Soil characteristics, such as natural water features and vegetation, are specified in the requirement class "LandFeature" as soil characteristics. Models of the land surface and subsurface layers are also included. Land improvements, such as fill construction or planting landscape materials are considered part of the site facilities in the "Facility" requirement class.

Part 2: Facility and Project

Facility

Facilities include collections of buildings and engineering structures and their associated site work. The facility requirement class provides only general support for the facilities themselves, allowing subsequent requirement classes to focus on specific types of parts that make up the facilities, such as road and rail. This requirement class is optional to allow for the condition in which not all information is related to the facility, for example, one containing only survey or land division information.

Different types of facility components are described in the "Facility Part Type" of the LandInfra model (Bridge, Building, Drainage, Environment, Railway, Road, Site, Tunnel, Wastewater, Water Distribution and Other). A LandInfra implementation is not required to support all these types. In fact, only road and rail are specified in detail in the initial version.

Project

The "Project" requirement class includes information about projects and their breakdown into project parts. To allow for the condition where none of the information in the LandInfra dataset is related to the project, this requirement class is optional.

Part 3: Alignment

An alignment is a positioning element that provides a linear referencing system for locating physical elements. The "Alignment" requirement class specifies how an axis is defined and used.

Part 4: Road and RoadCross

Road

The "Road" requirement class supports use cases in which a designer wishes to exchange the output of the design with someone who may use it for purposes other than completing the road design. Therefore, the "Road" requirement class includes several alternative methods for representing a design, for example, with 3D road elements, 3D chain lines (profile views, longitudinal breaklines and long sections) and 3D surfaces and layers, as well as combinations of these elements.

RoadCross

The "RoadCrossSection" requirement class extends the "Road" requirement class by adding the alternative 2D CrossSection ways of representing a design, as well as combinations of them.

Part 5: Railway

Like the "Road" requirement class, the "Railway" requirement class supports use cases in which a designer wishes to exchange the output of the railway design with someone who may use it for purposes other than the subsequent design. Therefore, the "Railway" requirement class covers design outputs such as 3D railway elements and track geometry, including superelevation.

Part 6: Survey, Equipment, Observations and SurveyResults

Survey

The "Survey" requirement class is the main class relating to surveys and provides a framework for information about observations, processes and their results collected during survey work. Due to the large number of classes, the "Survey" package has been divided into different parts: "Equipment", "Observations" and "SurveyResults".

Equipment

In the "Equipment" requirement class, all process and sensor information that was used to determine an observation is available.

Observations

The "Observations" requirement class is the package containing all the information about the raw observations and measurements taken or observed during the survey work.

SurveyResults

The requirement class "SurveyResults" contains the observed property of an element of interest. Using the sampling features of the Observation & Measurement (O&M) standard, the dependencies between the observational actions and the results are realized.

Part 7: LandDivision and Condominium

LandDivision

Land may be divided into land divisions, which may be public or private in nature. The former are administrative divisions and the latter are land interests. Both are specified in the "LandDivision" requirement class, although co-ownership rights in land are specified in a separate category, the "Condominium" requirement class.

Condominium

A condominium is the simultaneous ownership of real property that has been divided into private and common areas. The condominium requirements category includes information on condominium units, buildings and projects.

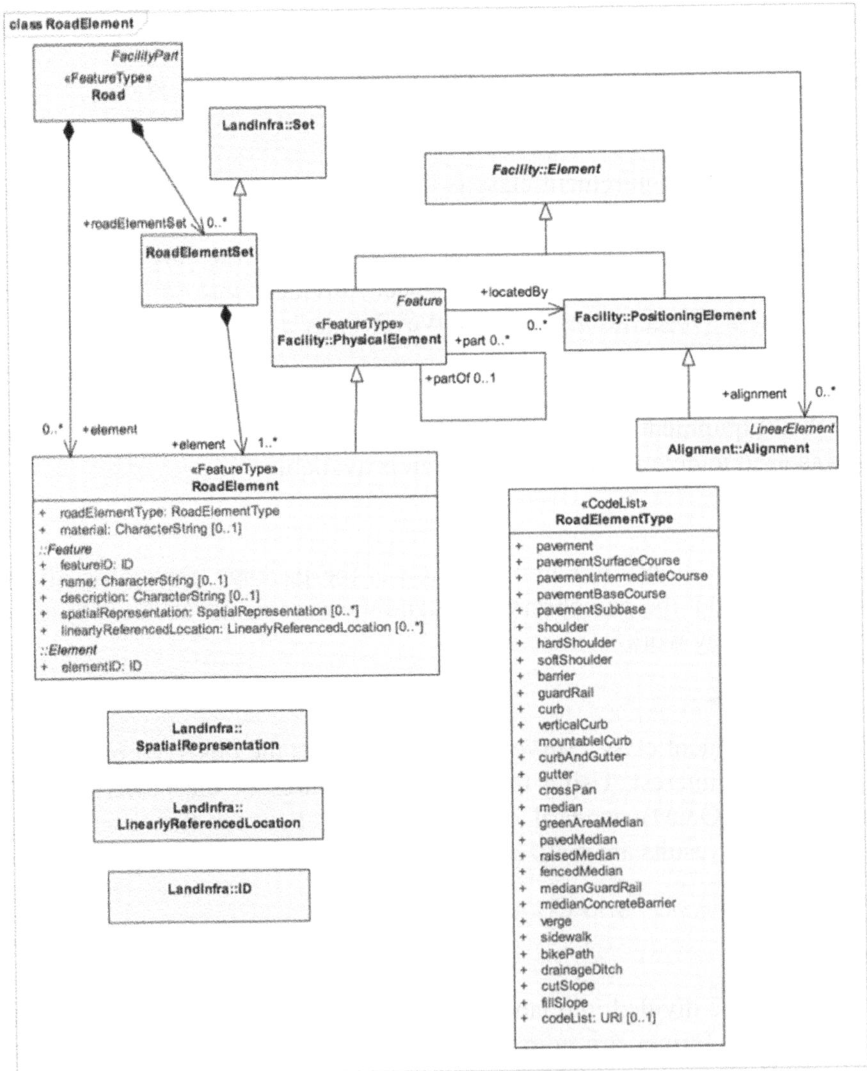

Figure 4.3. *RoadElement diagram (OGC 2016). For a color version of this figure, see www.iste.co.uk/hajji/BIM.zip*

We must note that the primary concern for the initial release of LandInfra is the determination of land ownership as it relates to infrastructure. For example, do I have sufficient rights-of-way to build a road or do I need to acquire additional property (or easements) from adjacent property owners? The legal/rights issues addressed by LADM are reduced in LandInfra to the needs necessary to define (and measure) the property boundary.

The LandInfra and CityGML standards overlap on the road and railroad classes. In LandInfra, Road FacilityPart is a collection of zero or more RoadElements (e.g. pavement layer, sidewalk and gutter), specified separately and/or in RoadElementSets (the same concept is applied to Railway FacilityPart). The RoadElement context diagram is shown in Figure 4.3.

For CityGML, the road is represented as a “TransportationComplex”, which is further subdivided into “TrafficAreas” and “AuxiliaryTrafficAreas”. TrafficAreas are those elements that are important in terms of traffic use, such as car lanes, pedestrian areas and bicycle paths. AuxiliaryTrafficAreas describe other road elements, such as curbs, middle lanes and green spaces.

TransportationComplex objects can be differentiated by theme using the subclasses Track, Road, Railway and Square. Each TransportationComplex has the attribute class, function and use.

4.3. Conclusion

From these data models, we conclude that each of them has its own specificities with some overlap. CityGML is the one that models the most urban objects above ground, with the ability to model underground infrastructure using Application Domain Extensions such as CityGML Utility Network ADE. IFC, with its latest versions, offers an object-oriented model of buildings and infrastructure (bridges, roads, railroads, ports and waterways) with the richest semantic model. LandInfra has the main objective of supporting land and civil engineering infrastructure (roads and railroads) with survey and land ownership information. All three standards support 3D geometries and represent objects in three dimensions.

4.4. References

buildingSMART (2020). IFC4x3 specification [Online]. Available at: https://technical.buildingsmart.org/.

Gröger, G., Kolbe, T.H., Nagel, C., Häfele, K.H. (2012). OGC City Geography Markup Language (CityGML) encoding standard. OpenGIS® encoding standard OGC 08-007rl, Open Geospatial Consortium, Inc.

OGC (2016). OGC® Land and infrastructure conceptual model standard (LandInfra) [Online]. Available at: http://docs.opengeospatial.org/is/15-111r1/15-111r1.html.

5

GeoBIM: Towards a Convergence of BIM and 3D GIS

5.1. Introduction

Intelligent and sustainable management of 3D urban space requires at least two scales of study. The first scale concerns the building as a reference urban object that is at the center of several urban issues. The second scale concerns the building's environment at the city or neighborhood level. Taking into account the richness and correlations between urban themes and also between urban objects (buildings, networks, vegetation, etc.), an integration of the two scales of analysis is necessary for the rigorous and relevant study of an urban space in all its dimensions. The result of this integration is the GeoBIM concept. It is a hybrid process combining information from BIM (at the building scale) and GIS (at the neighborhood, city, region, etc. scales).

BIM was initially developed for buildings and later extended to other urban objects such as infrastructure. With the increase in the number of applications requiring richer geometric and semantic data on urban space, BIM has taken on another urban dimension (multi-scale BIM), applying to a larger urban context by integrating several urban objects. Multi-scale BIM is not a simple extension of 3D modeling to the elements of an urban space but requires a management of the relationships between the objects of the digital model. Thus, an integration between BIM and CIM (City Information Modeling) through the concept of GeoBIM offers a framework of

---

Chapter written by Rafika HAJJI and Hassane JARAR OULIDI.

interoperability to implement applications that address both the “indoor” and “outdoor” contexts of buildings.

BIM and GIS are two domains that have been developed for different purposes and each address particular uses with distinct scales. CityGML and IFC, the basic standards for GIS and BIM, respectively, have different schemas in terms of geometry, topology and semantics. It is thus obvious that their integration faces conceptual and technical challenges.

The integration between BIM and GIS and, in particular, the interoperability between CityGML and IFC is a major research topic. Developments are directed towards finding solutions to match the schemas of the two models but are still not close to solving the problem of geometric and semantic interoperability between the two standards.

In this chapter we define the concept of GeoBIM, we highlight its advantages and we present them in relation to several themes. We then deal with the issue of integration between BIM and GIS as well as interoperability between IFC and CityGML models. At the end of the chapter, we present the latest research results in the field of GeoBIM.

5.2. The GeoBIM concept

The concept of GeoBIM emerged as a concept resulting from an integration between the two domains of BIM and GIS (Ellul *et al.* 2018) (Figure 5.1). The integration aims at a reciprocal enrichment of information from both domains and an extension of the scales of analysis (from BIM to GIS). GeoBIM finds a place in many applications in which the building must fit and interact with its urban context, such as the management of emergency services in buildings, indoor/outdoor navigation, environmental and energy management of buildings in an urban context, noise simulation and urban impact studies.

GeoBIM attracts a lot of interest within the scientific community. However, many challenges related to the integration of BIM and GIS are as yet unsolved. They are mainly due to the differences between the two domains, requiring the implementation of approaches and the development of conversion solutions between the two standards: IFC and CityGML.

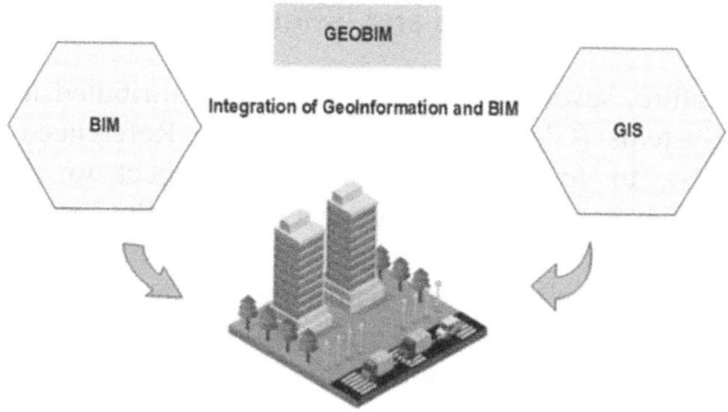

Figure 5.1. *The GeoBIM concept. For a color version of this figure, see www.iste.co.uk/hajji/BIM.zip*

5.3. Some applications of GeoBIM

A large number of applications have been developed around GeoBIM. In spite of the diversity of the domains addressed, the common point lies in the development of integration approaches, allowing us to take advantage of the geometric and semantic richness coming from the two domains of BIM and GIS to carry out thematic analyses in an urban space.

In the cadastral domain, several researchers have highlighted the need for the integration of BIM and 3D GIS in order to facilitate the registration and management of property rights (Dore and Murphy 2012, pp. 369–376). Notably, El-Mekawy *et al.* (2011, pp. 73–93) proposed an intermediate UBM (Unified Building Model) that combines IFC and CityGML semantic models for cadastral data representation. In the framework of the GeoBIM EuroSDR working group, Noardo *et al.* (2020a) proposed a GeoBIM-based workflow to automate the planning permit process in Europe. (Kang and Hong 2015) proposed an IFC and CityGML integration framework in the context of facility management and the smart city. Borrmann *et al.* (2015) proposed an IFC extension model to integrate many levels of detail representations with geometric dependency preservation in the context of tunnel modeling. Deng *et al.* (2016) developed a BIM and GIS-based platform for 3D noise mapping for urban planning and building interior design purposes in the ArcGIS environment. Rafiee *et al.* (2014) proposed a method based on georeferencing a BIM model for view and shading analyses in an urban context.

5.4. BIM and GIS: similarities and differences

In the literature, several definitions have been attributed to Geographic Information Systems (GIS), also called Spatially Referenced Information Systems (SIRS), to remove ambiguity with respect to software GIS. However, all definitions share geo-referenced data as the core data for a GIS application. Thus, a GIS can be defined as "an environment designed for the analysis and modeling of the spatial distribution of phenomena. It consists of a geographic database, a toolbox containing procedures for analysis, management, input and representation, and a user interface" (Collet 1992).

GIS completely supports two important aspects: the computerization of the problem and its spatial representation. A GIS places the problem in an informationally and representationally coherent spatial framework (Longley *et al.* 2015). The lifecycle of a GIS performs several functions, known as the "5A": Abstraction, Acquisition, Archiving, Analysis and Display (or *Abstraction*, *Acquisition*, *Archivage*, *Analyse* and *Affichage* in French) of spatially related data. 3D GIS is seen as a new generation of GIS that supports the various aspects of modeling, analysis, processing and representation of geographic information in 3D space through 3D geometric and topological structures.

BIM is a new paradigm in the field of construction, based on a collaborative process and working methods developed around a digital model. The BIM model is based on a parametric object-oriented modeling of a building that allows information to be attached to an object that encapsulates a set of attributes and relationships with other objects in the model (e.g. the relationship between a wall and a door, where the set is updated automatically to maintain the relationship). Unlike CAD (Computer Aided Design) models, BIM is based on a rich, consistent and continuously updated digital model.

Several similarities between BIM and GIS are referred to by Ellul *et al.* (2018). Both systems allow for the modeling of the built environment in 3D by integrating the interior and exterior features of that environment that can be managed in a DBMS (Database Management System). Both systems provide methods for documenting, editing, managing and visualizing spatial and non-spatial information at different scales and levels of detail.

The main differences between BIM and GIS lie in the purposes, study scales, spatial referencing of domains and modeling approaches (Figure 5.2).

Indeed, BIM was originally designed for specific needs in architecture, engineering and construction. GIS was instead designed to represent and analyze reality in a geographic coordinate system. GIS models the world as it exists while BIM develops a view to be projected into reality. BIM focuses on the building as an object of study, while GIS applies to a larger geographical area, including other urban objects. GIS adopts a global reference system on a national or regional scale, whereas BIM is linked to a local system.

The differences between BIM and GIS also lie in the geometric and semantic modeling approaches used in the relevant standards (IFC and CityGML). It is commonly accepted that BIM is more efficient in design and 3D modeling, as it allows us to represent complex objects with a high geometric accuracy. This is outside the capabilities of a GIS, which is more advanced in terms of spatial analysis.

Coming from different data sources, the two standards IFC and CityGML differ in terms of conceptualization of real-world objects. Indeed, IFC adopts a tree structure, according to which a building is modeled in several parts by including spaces (e.g. IfcBuildingStorey elements and IfcSpace elements), whereas CityGML adopts volumes for the representation of internal building elements such as rooms.

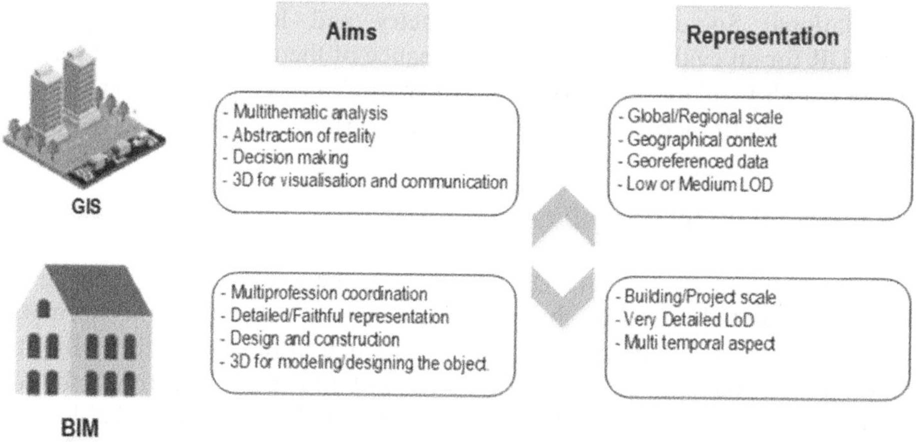

Figure 5.2. *Differences between BIM and GIS. For a color version of this figure, see www.iste.co.uk/hajji/BIM.zip*

5.5. BIM and GIS integration

In recent years, several research studies have been conducted to investigate the possibility of integrating the two domains of BIM and GIS under the concept of GeoBIM. Indeed, the integration of BIM and GIS meets several challenges. GeoBIM enables spatiotemporal analysis of geo-spatial problems at different levels of detail (Arroyo Ohori *et al.* 2018). GIS provides a context for extending building analysis to the urban scale and applying BIM methods to infrastructure projects while BIM data allows for the enrichment of urban models. Furthermore, this integration allows us to apply smart city concepts by adopting an integrated approach involving buildings, land, infrastructure and other urban objects.

Several research groups are addressing the issue of BIM and GIS integration, including the GeoBIM EuroSDR project that was launched in 2017 to study the feasibility of GIS and BIM integration and the IDBE (Integrated Digital Built Environment) group, which is a joint working group under buildingSMART International (bSI) and the Open Geospatial Consortium (OGC) that is working on integration in the area of GIS and BIM and interoperability between the IFC, CityGML and LandInfra (Land and Infrastructure Conceptual Model) standards.

The project "GeoBIM benchmark" was launched in 2019 by ISPRS (International Society for Photogrammetry and Remote Sensing) and EuroSDR for an evaluation of the performance of the tools developed for the interoperability of the conversion between IFC and CityGML and for the georeferencing of GeoBIM (Noardo *et al.* 2020b).

5.5.1. *Model interoperability*

In the general context of "Data Sciences", the exchange and sharing of data is essential. Given the heterogeneity of models and formats, interoperability is recognized as a major challenge. It represents the ability to guarantee that data generated by one user can be interpreted correctly by all other users (Shen *et al.* 2010). Data interoperability enables reliable and efficient information exchange: it is a prerequisite for effective system integration in a collaborative context.

Today, 3D models are used in many fields and cover multiple uses at different scales. However, the exchange of data between these models is a recurrent problem that hinders their dissemination and increases the cost of their production (Hajji 2013). Interoperability aims to integrate these models into a coherent framework that ensures fluidity in data transfer (El-Mekawy and Östman 2010) and their reuse outside the context in which they were created. Beyond a simple data overlay, the combination of multiple data sources imposes the management of geometric and semantic interoperability to align with the schema specifications of the final 3D model (Hajji 2013).

Interoperability in urban 3D modeling is the fundamental objective behind the development of the CityGML standard (El-Mekawy and Östman 2010). At the building scale, the AEC domain is undergoing a great evolution of practices towards a collaborative approach that requires the implementation of an interoperability framework. It has thus been necessary to standardize the models and data formats in BIM and to define methodological guidelines to standardize the processes. The goal is to eliminate or reduce time-consuming and error-prone manual interventions inherent to the operation or exchange of data between software and users. Standards developed by buildingSMART, such as IFC (Industry Foundation Classes), IFD (Information Framework for Dictionaries), MVD (Model View Definition) and IDM (Information Delivery Manual) improve the exchange of data between different BIM systems at the object level, thus minimizing the loss of information between several different applications.

The management of interoperability between IFC and CityGML is a prerequisite in the field of GeoBIM that has been addressed in several research studies. Several approaches have been adopted in this context, which we will analyze in the following section.

5.5.2. *Integration approaches*

The integration of BIM and GIS is at the heart of GeoBIM applications. Any integration approach must take into account the geometric and semantic mismatches between BIM and GIS standards (IFC and CityGML). There are two main families of approaches: (1) the unification or federation of schemas based on ontologies and the semantic web and (2) the schema matching approach (Figure 5.3).

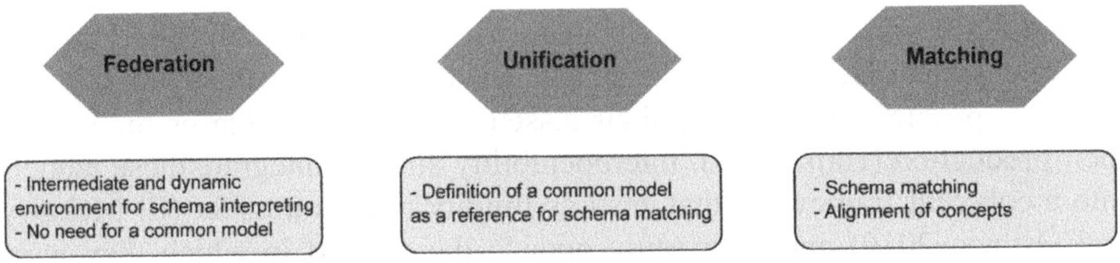

Figure 5.3. *Schema integration approaches*

Schema unification/federation approach

Schema unification is based on the definition of a common model serving as a reference for matching between the source schemas, whereas schema federation consists of creating an intermediate and dynamic environment to interpret instances of different schemas and to offer functionalities that apply to all data without having to define a common model. In this case, the use of ontologies is essential for the interpretation of semantic concepts from different domains.

Ontology and semantic web-based integration approaches operate at the conceptual level. In the presence of heterogeneous data, the use of an ontological integration approach relies on a semantic alignment of the concepts of two schemas based on a reference ontology describing the domain of study. The concepts of the schemas to be integrated can be aligned through this ontology (Figure 5.4). Such integration allows us to query the data using in queries that mask the underlying heterogeneities of the data (Cruz *et al.* 2004, pp. 51–66). In this context, Cruz *et al.* (2004, pp. 51–66) propose a semi-automatic tool for propagating alignments between different ontologies in the general context of geospatial data. El-Mekawy and Östman (2010) propose a reference ontology and a Unified Building Model (UBM) for the integration between IFC and CityGML.

The semantic web provides a framework for semantic interoperability between BIM and GIS through the Resource Description Framework (RDF) standard and the SPARQL web query language that manipulates and extracts data stored in the RDF format. Karan and Irizarry (2014) used these semantic web concepts in a BIM and GIS integration application in the context of facility management. An extension of IFC based on RDF vocabulary and ontologies is proposed by Beetz *et al.* (2014) in the context of port infrastructure management.

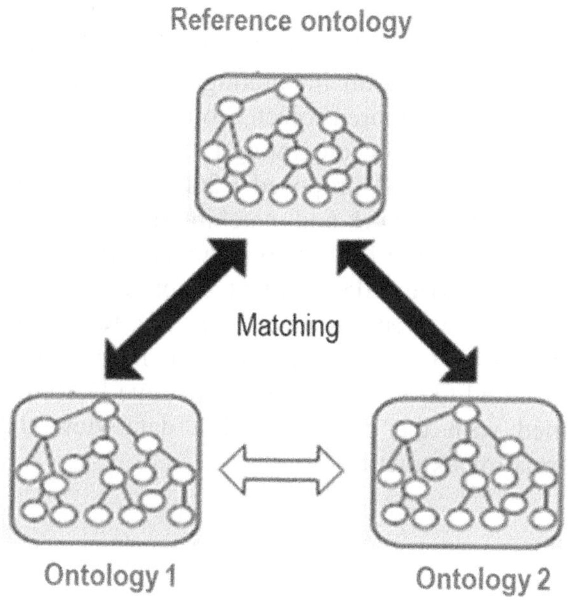

Figure 5.4. *Ontology-based approach*

Schema matching approach

The matching process can refer to the mapping between schemas, known as "schema matching" or between data, a process called "data matching" (Olteanu 2008). The two processes are not completely independent.

Schema matching uses several techniques. There are so-called "simple" techniques based on matching information directly from the schemas (class name, attributes, generalization relations, specialization, etc.) and techniques based on the declaration of Inter-Schema Matching Assertions (Sheeren *et al.* 2009) which are based on the identification of all correspondences between data schemas and between their instances at a semantic level (Read–Rewrite Process) (Figure 5.5).

Matching between IFC and CityGML schemas consists of aligning the semantic concepts of both standards. In the case where there is insufficient overlap between concepts, the use of Application Domain Extension (ADE) is an alternative to the "Read–Rewrite" process (Gilbert *et al.* 2020). Notably, de Laat and van Berlo (2011, pp. 211–225) developed and implemented a GeoBIM extension integrating IFC semantic concepts into CityGML. In the Virtual Singapore project, Stouffs *et al.* (2018) proposed an ADE application of CityGML for integrating IFC concepts that are not in the CityGML schema. Kang (2018) proposed B2GM (BIM-to-GIS conceptual mapping): a conceptual framework and matching mechanisms between BIM concepts and GIS based on the ISO 19166 standard. B2GM is built around three components: (1) BIM-to-GIS Element Mapping (B2G EM), defining the matching rules at the element level, (2) BIM-to-GIS LoD Mapping (B2G LM) for extracting LoD information from BIM with respect to the LoD concept in GIS and (3) BIM-to-GIS Perspective Definition (B2G PD), which is concerned with the definition of a data view that allows for multi-source information extraction.

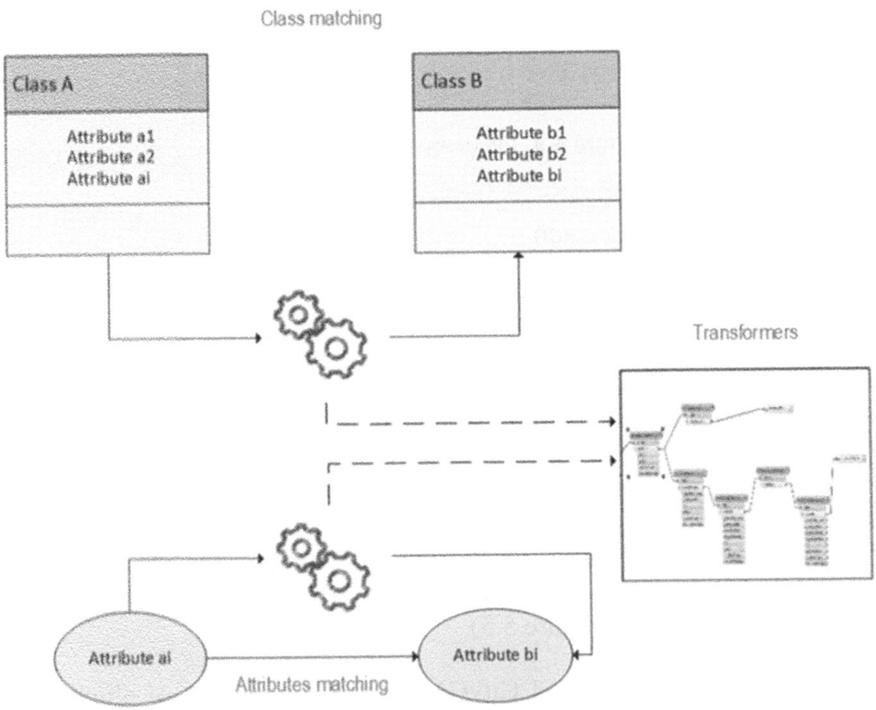

Figure 5.5. *Schema matching – "Read–Write" process. For a color version of this figure, see www.iste.co.uk/hajji/BIM.zip*

5.6. IFC and CityGML conversion

The integration of BIM and 3D GIS relies heavily on the exchange of data between the two schemas of IFC and CityGML. The IFC format allows us to exchange all or part of the object classes composing a BIM model to other GIS tools. For each object, IFC contains its geometry and the information attached to it. IFC geometry uses CSG modeling with volumetric parametric primitives representing the structural components of buildings. However, the CityGML standard uses surface representations (B-REP). This paradigm creates a combinatorial complexity in the transformation.

From a technical point of view, the conversion between CityGML and IFC can use ETL (Extract, Transform, Load) tools to transform BIM data into GIS data and vice versa. However, ETL tools such as FME (Feature Management Engine) are not close to solving all the geometric and semantic mismatches between IFC and CityGML and often present opaque interfaces that do not give the user the possibility of controlling the conversion. Moreover, these tools assume that the two models to be matched are well-structured according to the IFC and CityGML schemas.

The integration between IFC and CityGML has been the subject of several research studies (Donkers *et al.* 2016; Jusuf *et al.* 2017; Floros *et al.* 2018; Stouffs *et al.* 2018). Notably, Isikdag and Zlatanova (2009, pp. 79–97) proposed a framework for generating CityGML objects at multiple levels of detail from IFC objects. Donkers *et al.* (2016) developed a process for automatic conversion from IFC to CityGML LoD3.

Despite the developments, there is still not a perfect solution for conversion between IFC and CityGML and many problems identified in this conversion are yet to be solved (Kumar *et al.* 2019). A taxonomy of different geometric and semantic type errors from IFC-CityGML conversion is reported in Biljecki and Tauscher (2019).

5.7. BIM georeferencing

BIM has been developed as a new process in construction engineering to ensure, in particular, the co-design and collaborative monitoring of a construction project. Thus, the BIM model is linked to a local system without any requirement of the project's geographical referencing. As a logical consequence, georeferencing of mock-ups has never been a concern in the BIM process, and IFC entities to store this information have only been added in the latest versions (Noardo *et al.* 2020b).

The georeferencing of BIM plays a critical role in the insertion of a BIM project in its geographical context. To meet this need, the latest versions of the IFC standard propose classes for the georeferencing of BIM. Mainly, the entities "IfcSite" and "IfcGeometricRepresentationContext" allow us to store information about the spatial reference of the BIM. The "IfcSite" entity provides information on the geographical extent of the project through the attributes "RefLatitude", "RefLongitude" and "RefElevation", while the entity "IfcGeometricRepresentationContext" is used to define the spatial coordinates of a BIM model.

IFC defines several levels of georeferencing (LoGeoRef): LoGeoRef10, LoGeoRef20, LoGeoRef30, LoGeoRef40 and LoGeoRef50 (Figure 5.6).

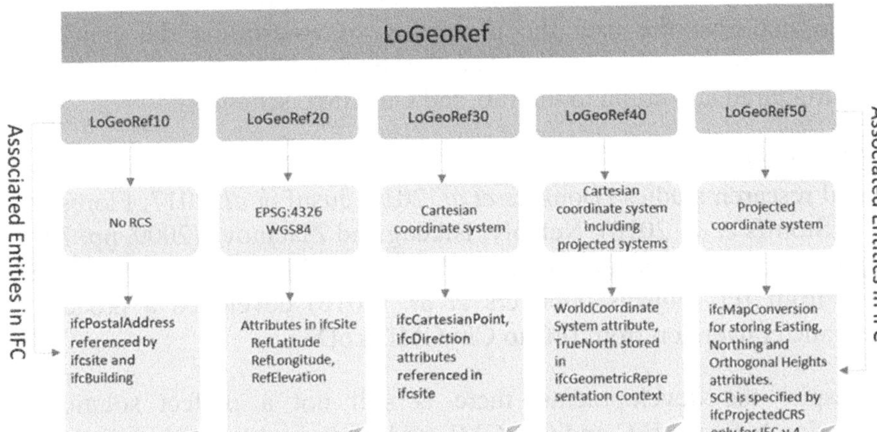

Figure 5.6. *Georeferencing options in IFC through "LoGeoRef" (inspired by Noardo* et al. *2020)*

Given the optional aspect of georeferencing, classes dedicated to georeferencing BIM models are often not complete. Software editors are thus called upon to implement LoGeoRef levels in georeferencing tools (Noardo *et al.* 2020b). As an alternative, some solutions have been developed in the literature. Notably, Diakité (2018) proposes a web interface to georeference BIM mockups in Revit using the Cesium platform. Diakité and Zlatanova (2020) develop an automatic approach for georeferencing a BIM in a GIS environment based on building footprints.

5.8. Conclusion

GIS and BIM are two systems designed for different purposes, and subsequently different in their models, formats and tools, each performing one aspect of 3D modeling at two different spatial scales. Their use in integrated approaches under the concept of GeoBIM requires the implementation of standardized approaches to ensure the interoperability of data and exchanges.

Given the technical challenges of integrating the BIM and GIS domains, stakeholders in both domains need to work closely together to establish a standard for GeoBIM. Some developments are underway. Version 3.0 of the CityGML standard (see https://github.com/opengeospatial/CityGML-3.0CM) is based on a new structuring of the core schema and application schemas and proposes the addition of entities allowing reconciliation between the BIM and GIS domains. The addition of a new concept of space will also allow the integration of new classes such as "BuildingConstructiveElement" to store detailed information on the elements of a BIM model in a CityGML file.

5.9. References

Arroyo Ohori, K., Diakité, A.A., Krijnen, T., Ledoux, H., Stoter, J. (2018). Processing BIM and GIS models in practice: Experiences and recommendations from a GeoBIM project in the Netherlands. *ISPRS International Journal of Geo-Information*, 7(8), 311.

Atazadeh, B., Rajabifard, A., Zhang, Y., Barzegar, M. (2019). Querying 3D cadastral information from BIM models. *ISPRS International Journal of Geo-Information*, 8(8), 329.

Beetz, J., van den Braak, W., Botter, R., Zlatanova, S., de Laat, R. (2014). Interoperable data models for infrastructural artefacts – A novel IFC extension method using RDF vocabularies exemplified with quay wall structures for harbors. *eWork and eBusiness in Architecture, Engineering and Construction*, 135–140.

Biljecki, F. and Tauscher, H. (2019). Quality of BIM–GIS conversion. *ISPRS Annals of Photogrammetry, Remote Sensing & Spatial Information Sciences*, 4.

Borrmann, A., Kolbe, T.H., Donaubauer, A., Steuer, H., Jubierre, J.R., Flurl, M. (2015). Multi-scale geometric-semantic modeling of shield tunnels for GIS and BIM applications. *Computer-Aided Civil and Infrastructure Engineering*, 30(4), 263–281.

Collet, C. (1992). *Systèmes d'information géographique en mode image*. Presses polytechniques et universitaires romandes, Lausanne.

Cruz, I.F., Sunna, W., Chaudhry, A. (2004). Semi-automatic ontology alignment for geospatial data integration. *Geographic Information Science*. Springer, Berlin, Heidelberg.

Deng, Y., Cheng, J.C.P., Anumba, C. (2016). A framework for 3D traffic noise mapping using data from BIM and GIS integration. *Structure and Infrastructure Engineering*, 12(10), 1267–1280.

Diakité, A.A. (2018). About the geo-referencing of BIM models. *Proceedings of 3D Geoinfo Conference*, Delft, 1–2 October.

Diakité, A.A. and Zlatanova, S. (2020). Automatic geo-referencing of BIM in GIS environments using building footprints. *Computers, Environment and Urban Systems*, 80, 101453.

Donkers, S., Ledoux, H., Zhao, J., Stoter, J. (2016). Automatic conversion of IFC datasets to geometrically and semantically correct CityGML LOD3 buildings. *Transactions in GIS*, 20(4), 547–569.

Dore, C. and Murphy, M. (2012). Integration of historic building information modeling (HBIM) and 3D GIS for recording and managing. In *2012 18th International Conference on Virtual Systems and Multimedia*, Guidi, G. (ed.). IEEE, Piscataway, NJ.

Ellul, C.D., Stoter, J., Harrie, L., Shariat, M., Behan, A., Pla, M. (2018). Investigating the state of play of GeoBIM across Europe. *International Archives of the Photogrammetry, Remote Sensing and Spatial Information Sciences*, 42(4/W10), 19–26.

El-Mekawy, M. and Östman, A. (2010). Semantic mapping: An ontology engineering method for integrating building models in IFC and CityGML. *3rd ISDE Digital Earth Summit*, Nessebar, 12–14 June.

El-Mekawy, M., Östman, A., Shahzad, K. (2011). Towards interoperating CityGML and IFC building models: A unified model based approach. In *Advances in 3D Geo-Information Sciences*, Heinrich Kolbe, T., König, G., Nagel, C. (eds). Springer, Berlin, Heidelberg.

El-Mekawy, M., Paasch, J., Paulsson, J. (2014). Integration of 3D cadastre, 3D property formation and BIM in Sweden. *4th International Workshop on 3D Cadastres*, Dubai, 9–11 November.

Floros, G.S., Ellul, C., Dimopoulou, E. (2018). Investigating interoperability capabilities between IFC and CityGML LOD 4 – Retaining semantic information. *International Archives of the Photogrammetry, Remote Sensing and Spatial Information Sciences*, XLII-4/W10, 33–40.

Gilbert, T., Rönsdorf, C., Plume, J., Simmons, S., Nisbet, N., Gruler, H.C., Kolbe, T.H., van Berlo, L., Mercer, A. (2020). *Built Environment Data Standards and their Integration: An Analysis of IFC, CityGML and LandInfra*. OGC Document. Open Geospatial Consortium, Wayland, and buildingSMART International, London.

Gotlib, D. and Karabin, M. (2017). Integration of models of building interiors with cadastral data. *Reports on Geodesy and Geoinformatics*, 104.

Hajji, R. (2013). Vers une méthode de conception de SIG 3D collaboratif. PhD Thesis, University of Liège.

Ibraheem, A.T., Hassan, H.A., Al-Husain, M.H.A. (2012). Integrating ACAD with GIS for civil engineering applications. *Journal of Software Engineering and Applications*, 5(3), 138–146.

Isikdag, U. and Zlatanova, S. (2009). Towards defining a framework for automatic generation of buildings in CityGML using building information models. In *3D Geo-Information Sciences*, Lee, J. and Zlatanova, S. (eds). Springer, Berlin.

Jusuf, S.K., Mousseau, B., Godfroid, G., Soh Jin Hui, V. (2017). Integrated modeling of CityGML and IFC for city/neighborhood development for urban microclimates analysis. *Energy Procedia*, 122, 145–150.

Kang, T.W. (2018). Development of a conceptual mapping standard to link building and geospatial information. *ISPRS International Journal of Geo-Information*, 7(5), 162.

Kang, T.W. and Hong, C.H. (2015). A study on software architecture for effective BIM/GIS-based facility management data integration. *Automation in Construction*, 54, 25–38.

Karan, E.P. and Irizarry, J. (2014). Developing a spatial data framework for facility management supply chains. *Construction Research Congress 2014: Construction in a Global Network*, 2355–2364, Atlanta, GA, 19–21 May.

Kumar, K., Labetski, A., Ohori, K.A., Ledoux, H., Stoter, J. (2019). Harmonising the OGC standards for the built environment: A CityGML extension for LandInfra. *ISPRS International Journal of Geo-Information*, 8(6), 246.

de Laat, R. and van Berlo, L. (2011). Integration of BIM and GIS: The development of the CityGML GeoBIM extension. In *Advances in 3D Geo-Information Sciences*, Kolbe, T., König, G., Nagel, C. (eds). Springer, Berlin, Heidelberg.

Longley, P.A., Goodchild, M.F., Maguire, D.J., Rhind, D.W. (2015). *Geographic Information Science and Systems*. John Wiley & Sons, New York.

Noardo, F., Ellul, C., Harrie, L., Overland, I., Shariat, M., Arroyo Ohori, K., Stoter, J. (2020a). Opportunities and challenges for GeoBIM in Europe: Developing a building permits use-case to raise awareness and examine technical interoperability challenges. *Journal of Spatial Science*, 65(2), 209–233.

Noardo, F., Harrie, L., Arroyo Ohori, K., Biljecki, F., Ellul, C., Krijnen, T., Eriksson, H., Guler, D., Hintz, D., Jadidi, M.A., Pla, M., Sanchez, S., Soini, V.-P., Stouffs, R., Tekavec, J., Stoter, J. (2020b). Tools for BIM-GIS integration (IFC georeferencing and conversions): Results from the GeoBIM benchmark 2019. *ISPRS International Journal of Geo-Information*, 9(9), 502.

Oldfield, J., van Oosterom, P.J.M., Quak, W., van der Veen, J., Beetz, J. (2016). Can data from BIMs be used as input for a 3D cadastre? *5th International FIG 3D Cadastre Workshop*, 199–214, Athens, 18–20 October.

Olteanu, A.M. (2008). Fusion de connaissances imparfaites pour l'appariement de données géographiques : proposition d'une approche s'appuyant sur la théorie des fonctions de croyance. PhD Thesis, Université Paris-Est.

Rafiee, A., Dias, E., Fruijtier, S., Scholten, H. (2014). From BIM to Geo-analysis: View coverage and shadow analysis by BIM/GIS integration. *Procedia Environmental Sciences*, 22, 397–402.

Sheeren, D., Mustière, S., Zucker, J.D. (2009). A data-mining approach for assessing consistency between multiple representations in spatial databases. *International Journal of Geographical Information Science*, 23(8), 961–992.

Shen, W., Hao, Q., Mak, H., Neelamkavil, J., Xie, H., Dickinson, J., Thomas, R., Pardasani, A., Xue, H. (2010). Systems integration and collaboration in architecture, engineering, construction, and facilities management: A review. *Advanced Engineering Informatics*, 24(2), 196–207.

Shojaei, D., Rajabifard, A., Kalantari, M., Bishop, I.D., Aien, A. (2014). Design and development of a web-based 3D cadastral visualisation prototype. *International Journal of Digital Earth*, 8(7), 538–557.

Stouffs, R., Tauscher, H., Biljecki, F. (2018). Achieving complete and near-lossless conversion from IFC to CityGML. *ISPRS International Journal of Geo-Information*, 7(9), 355.

Sun, J., Mi, S., Olsson, P.-O., Paulsson, J., Harrie, L. (2019). Utilizing BIM and GIS for representation and visualization of 3D cadastre. *ISPRS International Journal of Geo-Information*, 8(11), 503.

Tae-Wook, K., Seung-Hwa, P., Chang-Hee, H. (2016). BIM/GIS-based data integration framework for facility management. *GEOProcessing 2016: The Eighth International Conference on Advanced Geographic Information Systems, Applications, and Services*, 100–105, Venice, 24–28 April.

6

BIM and 3D GIS Integration for Real Estate Valuation

6.1. Introduction

In recent decades, accelerating rates of urbanization and land scarcity have led to the expansion of vertical and increasingly complex urban forms. This spatial configuration presents a major challenge for the accurate determination of property values.

Estimating the real estate value of a property is the process of identifying and quantifying each determinant of that value (RICS 2020). The volumetric nature of a property dictates the use of 3D models for accurately determining real estate value. Throughout the lifecycle of a residential asset, several stakeholders (investors, developers, decision-makers, tax authorities, banks and insurance companies, etc.) require the valuation of these properties as a main indicator for any future action (acquisition, construction, transfer of ownership, value adjustment, etc.).

Studies have recently shown that 3D modeling of the variables of a property value results in more accurate estimates (Işıkdağ *et al.* 2015). A number of models have been developed for this purpose. Among them is the International Valuation Information Model, which is an extension of the Land Administration Domain Model (LADM) from ISO 19152:2012 (Kara *et al.* 2018, 2020).

Chapter written by Siham EL YAMANI and Rafika HAJJI.

Building Information Modeling (BIM) is one of the most recent paradigms in the real estate industry, which is becoming increasingly popular in construction engineering, but whose use also extends to several topics associated with "Smart Building", including energy efficiency, noise propagation, urban regulation, real estate valuation and so on. 3D Geographic Information Systems (3D GIS), on the other hand, have the potential for geospatial analysis and 3D visualization, which can simulate the impact of several external factors on the real estate value of a property in a 3D urban environment.

BIM and 3D GIS have different purposes, but their integration offers great potential in the production of digital mock-ups at the urban scale. Such integration will provide a 3D information base to undertake real estate value simulations in both an indoor and outdoor context of buildings. Research in this area is still in its infancy and investigations in the literature are far from proposing a practical framework for real estate valuation that takes into account the combination of several factors and their dynamic simulation at an urban scale.

In the following sections, the concept and approaches of real estate valuation are presented, the main potentialities and applications of BIM and 3D GIS in real estate valuation are discussed, and the challenges that come with integrating BIM and 3D GIS in the accurate determination of the value of a property are analyzed. Also, we illustrate, through examples of uses, how BIM and 3D GIS can contribute individually or jointly in 3D modeling and simulations of variables in real estate valuation.

6.2. Real estate valuation: concepts, approaches and standards

The residential property market plays a prominent role in economic development. Thus, any inaccurate estimate of property value will have deplorable consequences for future developments in the construction industry and real estate economy (Curran *et al.* 2018). Many urban cores may be subject to overvalued residential property estimates that make housing prices increasingly unaffordable.

6.2.1. *The concept of real estate valuation*

Real estate appraisal is a complex process of determining the best price for which a property can be exchanged (Pagourtzi *et al.* 2003). Much research has discussed the issue of the accuracy of real estate appraisals. Some of them assume that the reliability of a real estate valuation depends on the quality of the variables considered (Mooya 2016). The choice of a valuation approach and model are also key criteria in the accurate assessment of values. These variables differ from one appraisal to another depending on the type of stakeholder (real estate expert, tax authorities, etc.) and their opinion on the value, the object of the appraisal (transfer of property, future construction, renovation, real estate credit, etc.) and the estimation approaches used (comparison item, cost approach, hedonic approach, etc.). However, certain variables are common to all use cases, such as the location of the property.

The value of a residential property is closely linked to the determination and integration of the inherent characteristics of the residential property (location, surface, accessibility, noise, age, etc.). It is dependent on several factors that can be classified into internal and external factors. The internal factors include the surface area of the dwelling, the density of the building (ratio of the habitable surface area to the sum of the floor areas), the number of rooms, the elements of comfort (sunshine, ventilation, acoustic comfort, thermal comfort), the number of floors, and so on. The external characteristics include all the attributes to do with the micro- and macro-spatial location of the dwelling, allowing us to understand the dynamics of the place (accessibility of transport, social atmosphere, etc.): the geographical location, social environment and physical environment. These variables can also be classified into subjective and objective variables (Chan and Abidoye 2019). The so-called objective variables are usually directly quantified (accessibility and structure variables). Subjective variables are determined qualitatively, either through the appraiser's assessment as a connoisseur of the market, or through a survey that attests to the preferences of potential buyers in a small market (Zhang 2019).

6.2.2. *Real estate valuation approaches*

There are several real estate valuation methods. There are so-called traditional and advanced methods (Pagourtzi *et al.* 2003). The traditional methods consider a residential property as a homogeneous entity comparable to other similar entities in terms of selling price, income and cost. These are the comparative approach (Adetiloye and Eke 2014), the income approach and the cost approach also known as the developer balance sheet. However, the advanced methods are based on automatic modeling techniques and require a significant history of real estate transaction data in order to estimate the value of the property (Monson 2009). These include Artificial Neural Network (ANN) techniques (Chan and Abidoye 2019), hedonic price models coupled with geospatial analysis methods, autoregressive–moving-average (ARMA) models (Cohen and Coughlin 2008), etc.

Recent research has revealed the inability of traditional approaches to evaluate residential property objectively and accurately (Mooya 2016). The lack of available market data and the absence of models that consider the complex and heterogeneous nature of residential property leave real estate valuation practices subjective and non-transparent.

6.2.3. *Real estate valuation norms and standards*

There are several regional and international associations and professional bodies whose role is to establish real estate valuation standards. Some of these standards include: the European Valuation Standards (EVS – The Blue Book) (TEGoVA 2016), the International Valuation Standards (IVSC 2016) and a set of standards published by the International Association of Assessing Officers (IAAO). These standards focus more on the procedural aspects of real estate appraisal and the definition of so-called conventional methods to define the practical exercise of real estate appraisals. For example, the International Valuation Standards (IVS) set forth the broad principles of valuation.

The Land Administration Domain Model (LADM) is an international standard that defines data modeling for land administration. LADM is a conceptual model that meets the guidelines of ISO 19152:2012. This model provides an extensible basis for the development of other data models related to property management, land value and land use. The LADM Valuation Information Model provides detailed information on legal, geometric and physical characteristics for property valuation units (Kara *et al.* 2018).

It provides a basic model for managing data about stakeholders in appraisal practices, appraisal entities and their characteristics. Note that this model focuses only on administrative assessments related to property tax assessments and excludes other public and private sector assessment activities.

6.3. BIM and 3D GIS for real estate valuation

The literature identifies several challenges for real estate valuation (Meins *et al.* 2010), among which we mainly mention: the management of uncertainty in the value estimates of numerous parameters related to real estate value as well as the lack of transparency and the subjectivity of the expert. The resolution of these challenges requires the use of numerical methods that allow the integration of different variables and their simulation within the same model.

As a result of research into the identification of three-dimensional variables that influence the determination of residential value, we propose the following classification:

– **Variables based on 3D data sources:** 2D or non-spatial data that are potentially extracted from 3D models (e.g. 2D surface from a 3D model).

– **Variables requiring 3D modeling (quantification):** variables whose values can be more accurately determined through 3D modeling and simulation.

– **Variables based on a 3D environment (qualification):** 3D modeling is not intended to overcome the subjective assessment of certain variables but rather to provide decision-makers and evaluators with a 3D view to facilitate their evaluations.

The emergence of building information acquisition and modeling technologies, and the rich semantic and geometric characteristics of its building elements (Eastman *et al.* 2011), have motivated researchers to exploit the potential of BIM models in property management, taxation and valuation. BIM provides a digital building information model based on object-oriented modeling rich in geometric, semantic and parametric building data, from which views and data appropriate for the needs of different users can be extracted and analyzed to produce information that can be used to make decisions. Through this type of modeling, BIM can address several real estate assessment challenges:

– BIM allows one to integrate in the same model several parameters related to the characteristics and behavior of the building to have an accurate estimate of the value of the property, and to take into account the interactions between the various parameters.

– BIM ensures a dynamic simulation of the construction cost, which is a very significant factor in the real estate evaluation.

– BIM makes it easy to more accurately determine some parameters related to certain internal characteristics of the building, such as sunlight, lighting, thermal comfort and acoustic comfort.

– BIM supports all the simulations related to the building envelope as well as the different elements to analyze the impact of a certain geometrical configuration on the real estate value of a property.

Atazadeh *et al.* (2017) discussed the possibility of using BIM to model 3D land administration. Their work highlighted the challenges of the current land administration process and the importance of integrating 3D information. Kara *et al.* (2018) identified the type of analysis that can be used for land valuation in the context of the land administration domain model. They also discussed the importance of integrating 3D analysis into land valuation and conducted analyses using a dataset from the Netherlands.

BIM models take advantage of the potential of the IFC standard in the modeling of the internal elements of the building (volumes, cost and energy simulation). However, it remains limited to applications in the field of architecture and construction. Several studies have concluded that it is possible to develop IFC extensions adapted to several domains, namely applications in terms of 3D property management. Atazadeh *et al.* (2017) use the spatial element "IfcSpace" to manage 3D legal property objects and to propose a prototype that enriches BIM/IFC models with legal semantic information.

Beyond the building scale, 3D GIS adds another dimension of 3D visualization and spatial analysis, to better estimate the value of a property in relation to several parameters characterizing the external environment of the building, which can be physical, environmental, social and so on. Thus, the BIM/IFC modeling approach can be extended to the urban scale, aiming to combine its strong semantic capabilities with the spatial analysis tools offered by GIS systems in order to assist the real estate value simulation process (Plume and Mitchell 2011).

In real estate appraisal, the purpose of 3D visualization is to visualize the input elements for an appraisal and let the user (e.g. the appraiser) subjectively qualify the value of the composite elements of the value according to their expertise and knowledge of the area and the real estate market. This subjectivity can significantly affect the accuracy and impartiality of the results of a real estate valuation. In some cases, visualization provides additional information to the appraiser by allowing them to take into consideration certain configurations, such as the presence of obstructing buildings nearby and 3D intervisibilities. While visualization is useful for the appraiser to take into account certain external factors in the space around the building in the value simulation, other elements necessary for real estate appraisal are derived from geospatial analysis. This is the case, for example, for the simulation of the building's energy performance.

The use cases for 3D city models are numerous and cover several domains (Biljecki *et al.* 2018): energy simulations, noise modeling, 3D cadastral, multi-temporal analysis, flow analysis and so on. Real estate valuation can also benefit from 3D GIS models. In a broader perspective and with the advent of urban models, the insertion of the building in an urban context will make it possible to easily verify the respect of the urbanistic potentialities in front of the proposed conceptual design of the real estate, and make it possible to readjust the elements of future valuation of the real estate before its construction as well as a prior evaluation and adjustment of its value. Işıkdağ *et al.* (2015) demonstrated that 3D GIS models could be used for the valuation of existing properties. Practically, their research aimed to identify the necessary information elements for the taxation value of real estate.

6.4. BIM-3D GIS integration: a new paradigm for real estate valuation

The possibilities to access and use 3D geographic information in very specific application areas have increased exponentially in recent years. However, many countries still do not have 3D city models, or do not allow access to the geographic information related to these models and rely on 3D visualization applications. Specifically, real estate appraisal adds another challenge, that of accessing transaction information and value estimation histories for each vertically arranged property. While this information can be accessed in some cases in administrative registration databases, it is incomplete and only reflects average values for a specific area, which is limited to taxation applications.

The 3D geometric data of buildings includes their height, floor area, roof area and orientation. This type of data is still underutilized in real estate price analysis, but research to date has shown promising results in improving predictions. Also, studies have shown that noise in a neighborhood is influenced by building size. The age of a building, an important feature of many hedonic price models, can also be estimated using 3D data. 3D data can be used to digitally represent views of apartments, taking into account the height of neighboring buildings and orientation.

BIM and 3D GIS integration is prominent in many applications. The previous section highlighted their promising potential in real estate valuation. Several 3D factors that make up real estate price and value have been the subject of recent research. These relate to the physical characteristics of the property as well as the environmental factors of its immediate vicinity. The integration of BIM and 3D GIS makes it possible to account for these two types of parameters (Figure 6.1) as well as the relations that can exist between them.

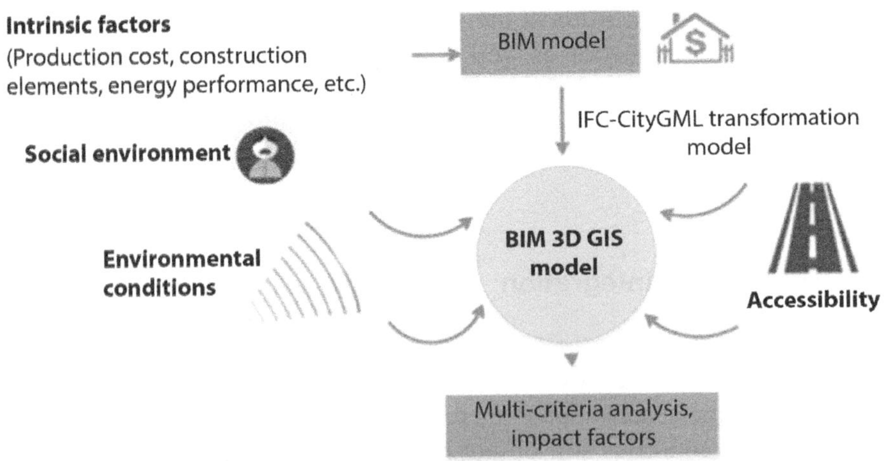

Figure 6.1. *BIM and 3D GIS integration for real estate valuation. For a color version of this figure, see www.iste.co.uk/hajji/BIM.zip*

BIM-3D GIS integration is a complex process that presents major technical challenges, mainly due to the heterogeneities between the two information models in terms of spatial scale, level of granularity and detail (LoD), geometry representation methods, storage and access methods, as well as semantic differences between the models.

In the context of urban applications, the integration of the BIM urban model in a 3D GIS environment is an embryonic niche, particularly in real estate appraisal, for which there are mainly three integration approaches:

– Application level: real estate valuation use cases at this level are limited to visualizing the BIM model for cost estimation in a 3D GIS environment, for example (Arcuri *et al.* 2020).

– Process level: this level consists of setting up an architecture to integrate BIM and 3D GIS into a workflow where they are simultaneously operational but remain separate. This approach is flexible, but presents several challenges in terms of interoperability. This is the most recommended level for real estate appraisals.

– Data level: this is a more flexible level of data integration where one of the two models (BIM or GIS) is extended through its standard to accommodate the data and elements of the other model. Another more advanced level of extension is the development of a meta-model that mediates between the two models at a high conceptual level. The GeoBIM extension by de Laat and van Berlo (2011) is an example of extending CityGML with BIM elements.

The challenge of interoperability between the two BIM and 3D GIS models does not allow us to take full advantage of its potential in real estate valuation. Most research is limited to the first two levels of integration, which results in significant data loss during this process. However, the third level is the most suitable in terms of accuracy, performance and applicability in real estate appraisal.

6.5. Examples of BIM and 3D GIS simulations for real estate valuation

Table 6.1 below summarizes the different use cases of BIM and 3D GIS for real estate valuation, which are analyzed and discussed in the following sections.

Use cases	Example of application in real estate value	Model	References
Simulation and quantification of internal factors of a property	Construction cost estimates Energy simulation	BIM	El Yamani *et al.* (2019), Morano *et al.* (2019) and Arcuri *et al.* (2020)
Identification of comparable units	Comparative approach	BIM/3D GIS	El Yamani *et al.* (2019) and Couto *et al.* (2021, pp. 15–19)
3D visibility analysis	Evaluation model	3D GIS	Yu *et al.* (2007) and Kara *et al.* (2018)
3D cadastral model	Real estate tax value Real estate price benchmark Real estate transactions	BIM/3D GIS	Işıkdağ *et al.* (2015) and Kara *et al.* (2020)

Table 6.1. *Potential use cases for BIM and 3D GIS in real estate assessment*

6.5.1. *Simulation of the internal factors of a property*

BIM is a detailed digital representation of geometric, semantic and parametric data of the building. It is a model that is becoming increasingly used in several fields, notably in the energy efficiency of buildings, by analyzing sunlight, illumination and ventilation. In addition to simulations of the structural and architectural elements of buildings, BIM simulations in the field of construction economics also have an important contribution for a dynamic calculation of the time and cost of construction to save budgets and optimize deadlines. BIM also allows a micro-environmental analysis of the property to be evaluated, to be able to identify the factors on which its value depends, namely indoor air quality, sunlight, acoustic comfort, humidity, ventilation, and so on (Figure 6.2). This is done by performing analyses related to the internal environmental parameters of the building.

BIM software allows an automatic and precise extraction of the elements determining the value of a property through quantity extractions (take off quantities, calculation of surfaces and volumes, etc.). The construction cost approach calculates the construction costs of a building and adds the land costs to determine the value of a property. To calculate the exact construction costs of a property, the BIM model must be characterized by a high semantic accuracy on the building data.

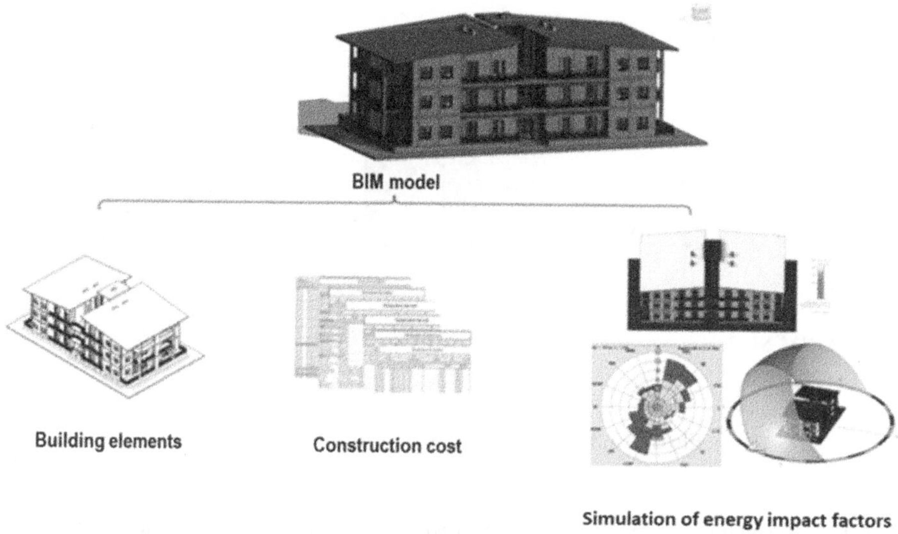

Figure 6.2. *Internal factors for assessing the value of a property. For a color version of this figure, see www.iste.co.uk/hajji/BIM.zip*

Generally, the cost estimation can be broken down into two steps. The first step consists of quantity take-off from an IFC file. The second step involves the determination of elementary prices (Figure 6.3). Three cost estimation options can be distinguished (Eastman *et al.* 2011): (1) exporting the quantified building elements to a cost estimation software; (2) using a BIM tool (Quantity take-off); and (3) linking the BIM tool directly to the cost simulation software. Tools for estimating the cost from a BIM model include: Open BIM Cost estimate, BIM C, BIM OFFICE, BIM Estimate, Vico Office, BIM Vision, Innovaya, JustBIM, Cost X, Cost OS, Dprofiler, Autodesk Quantity Take-off and Autodesk Navisworks.

Cost estimation is strongly related to the quality, geometric accuracy and level of information of the model. In the absence of specific standards for cost estimation, the level of detailed specification for building structures generally refers to the Level of Development (LoD). The LoD clearly defines the elements that must be included in the model. LoD 350 appears to be the most optimal for cost estimation (Vitasek and Zak 2018).

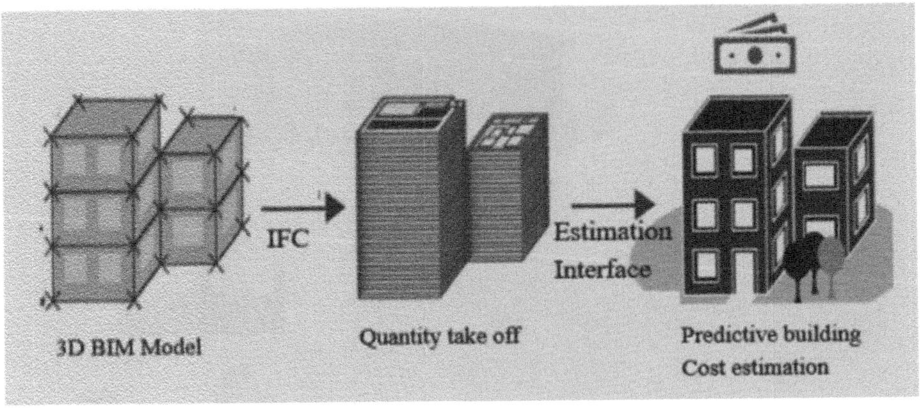

Figure 6.3. *Cost estimation from an IFC model*

Other internal parameters determining the estimation of the value of a property concern the thermal comfort within a building. Among these parameters, we can mention ventilation, internal noise, sunshine and so on. Indeed, air quality and ventilation contribute to the promotion of a healthy and comfortable environment. Acoustic analysis, on the other hand, through noise simulation, is very important to the comfort within a building and that should be taken into consideration when designing the form and choosing building materials (Christensen 2018). Sunlight analyses allow us to predict the energy consumption of the future building. Determining this factor is important for real estate valuation because it allows for improved occupant comfort and health. Analyses before the construction of the building make it possible to predict the optimal design to save electrical energy consumption. The simulations are translated into several factors: the apparent sun path, shading and lighting. These analyses are very useful in the pre-construction phase, as an adequate daylighting plan can reduce energy costs and avoid unwanted heat absorption in summer. The location and dimensions of openings (e.g. windows and doors) are decided based on the result of this analysis (Khan *et al.* 2019).

6.5.2. *Identification of comparable units of the property being appraised*

These applications primarily use previously transacted/leased or appraised properties as a comparable basis for determining the value of appraised properties. This enhanced sales comparison approach based on appraisal methods can use a 3D GIS to find comparable sales units based on various factors impacting the property value. Using a 3D city model to identify comparable sales units is primarily limited to visualization and leaves room for the subjectivity of the appraiser. Different spatial operations are used to calculate the factors mentioned above. The search for comparable properties in terms of physical, structural or environmental characteristics is mostly applicable to any real estate valuation use case:

– estimating the tax liability of a property to find comparable ones;

– estimating the price/value influence factors.

6.5.3. *3D visibility analysis*

Geospatial analysis through 3D GIS allows for the accurate identification of real estate value in a three-dimensional environment. The research focuses on the property visibility polygon as an influencing variable on property value. The objective of a 3D model is to analyze and quantify the visibility polygon of real estate objects on large surfaces. The added value is mainly in terms of efficiency by performing the analysis and visualizing the results in a 3D space. Without the 3D model, an observing agent would have to determine the visibility polygon of a building by passing through each floor and making measurements from different positions, while respecting the level of detail required to perform the 3D simulation. In practice, visibility modeling requires the availability of openings/facades for the 3D model, or even an intrinsic/extrinsic advanced quality. Such modeling will allow the following simulations:

– estimated exterior noise level at the front of the property at different floors;

– pollution level at the window/opening level;

– air quality at property openings;

– quality of view to observe from an aperture;

– rate of exposure to natural sunlight from outside and inside the property;

– proximity to various sources of noise/pollution.

6.5.4. *Valuation of tax property elements: 3D cadastre*

3D cadastral models are widely explored for the management and modeling of real estate registration rights. Thus, these models are introduced to estimate the factors that influence the average value of real estate according to the types of assets. This reference value is considered as a basis for taxation.

Real estate valuation is, in most countries, the basis for calculating taxes. The value depends on several factors, including physical characteristics (geometry, building materials, and other characteristics related to quality and occupant satisfaction) and legal factors (rights, restrictions and liabilities) (Işıkdağ 2015). Işıkdağ *et al.* (2015) discuss the use of digital building models and cadastral models when determining real estate valuation information. However, real estate appraisal stakeholders are not yet taking advantage of the potential of BIM and 3D cadastre in real estate valuation (Işıkdağ *et al.* 2015)

Legal information modeling allows the creation of 3D legal objects. These illustrate the administrative limits of the property in volume. The fact that the space allowed for construction is larger than the size of the construction has a positive influence on its market value. In addition, other types of legal information can influence the value of a property (rights are a surplus value, easements a deficit value) (Işıkdağ *et al.* 2015).

The exploitation of the 3D cadastral models to estimate the basis for the tax value of real estate generally gives an average value per area. Thus, automated mass real estate valuation approaches are advocated for rather than individual property estimation. Automated mass appraisal combines large amounts of data to predict the average value of real estate in a specific zoning through 3D models (Tomić *et al.* 2012).

6.6. Conclusion

Real estate value estimation is a complex process that requires different data sources and can benefit from the contribution of BIM and 3D GIS. To our knowledge, there is not yet an application that takes into consideration the different use cases of valuation based on BIM and 3D GIS models.

The challenge of digital modeling methods through BIM and 3D GIS is to recover these data which are often poorly archived and dispersed, or not at all. The ideal situation would be to be able to gather all the documentation that is generated on the building so as to avoid redoing additional studies. The challenge is the development of a standardized digital format for the storage of data and documentation such as BIM and 3D GIS, which would allow the creation of a database on the building from its design to its operation.

Standardization allows one to structure the necessary elements for each valuation and application case, which comes from elements related to the scale of the real estate property (BIM based) and its 3D environment (3D GIS based) as well as transactional and previous valuation databases. All should be connected by well-defined geometric and semantic relationships to facilitate simulations in 3D property valuation models. However, standardization efforts are mainly focused on the use of 3D cadastral models and real estate transaction databases.

The enrichment of BIM and 3D GIS standards including concepts related to real estate applications will allow some technical constraints related to BIM and 3D GIS integration to be overcome. Recently, CityGML 3.0 adds a new concept of "space" and the possibility to include the modeling and simulation of 3D variables related to the indoor space of buildings (Tauscher 2020).

6.7. References

Adetiloye, K.A. and Eke, P.O. (2014). A review of real estate valuation and optimal pricing techniques. *Asian Economic and Financial Review*, 4(12), 1878–1893.

Arcuri, N., De Ruggiero, M., Salvo, F. (2020). Automated valuation methods through the cost approach in a BIM and GIS integration framework for smart city appraisals. *Sustainability*, 12(18), 7546.

Atazadeh, B., Kalantari, M., Rajabifard, A., Ho, S., Ngo, T. (2017). Building information modelling for high-rise land administration. University of Melbourne.

Biljecki, F., Kumar, K., Nagel, C. (2018). CityGML application domain extension (ADE): Overview of developments. *Open Geospatial Data, Software and Standards*, 3(1), 1–17.

Chan, A.P.C. and Abidoye, R.B. (2019). Advanced property valuation techniques and valuation accuracy: Deciphering the artificial neural network technique. *RELAND: International Journal of Real Estate and Land Planning*, 2, 1–9.

Christensen, J.K. (2018). Using BIM as a tool to decrease variation in quantity and optimize the workflow of generating a cost estimation. In *Building Information Modeling (BIM) 2: Further Exploration in Architecture, Engineering and Construction*, Kamari, A. and Kirkegaard, P.H. (eds). Aarhus University, Aarhus.

Cohen, J.P. and Coughlin, C.C. (2008). Spatial hedonic models of airport noise, proximity, and housing prices. *Journal of Regional Science*, 48, 859–878.

Couto, P., Silva, M.J.F., Salvado, F. (2021). *Real-Estate Valuation Based on BIM Methodology*. Springer, Cham.

Curran, M., Spillane, J., Clarke-Hagan, D. (2018). External stakeholders in urban construction development projects: Who are they and how are they engaged? In *34th Annual Conference*, Gorse, C. and Neilson, C.J. (eds). Association of Researches in Construction Management, Belfast.

Eastman, C.M., Eastman, C., Teicholz, P., Sacks, R., Liston, K. (2011). *BIM Handbook: A Guide to Building Information Modeling for Owners, Managers, Designers, Engineers and Contractors*. John Wiley & Sons, Hoboken, NJ.

El Yamani, S., Ettarid, M., Hajji, R. (2019). Building information modeling potential for an enhanced real estate valuation approach based on the hedonic method. *WIT Transactions on the Built Environment*, 192, 12.

Işıkdağ, Ü., Horhammer, M., Zlatanova, S., Kathmann, R., Van Oosterom, P.J.M. (2015). Utilizing 3D building and 3D cadastre geometries for better valuation of existing real estate. *Proceedings of the FIG Working Week 2015 'From the Wisdom of the Ages to the Challenges of Modern World'*, Sofia, Bulgaria, 17–21 May.

IVSC (2016). International Valuation Standards [Online]. Available at: https://www.ivsc.org/standards/international-valuation-standards/IVS.

Kara, A., van Oosterom, P.J., Çağdaş, V., Işıkdağ, Ü., Lemmen, C.H.J. (2018). 3D data for better property value estimation in the context of LADM valuation information model. *6th International FIG Workshop on 3D Cadastres 2018*, Delft, 2–4 October.

Kara, A., van Oosterom, P.J.M., Çağdaş, V., Işıkdağ, Ü., Lemmen, C. (2020). 3 dimensional data research for property valuation in the context of the LADM Valuation Information Model. *Land Use Policy*, 98, 104179.

Khan, A. and Ghadg, A.N. (2019). Building Information Modelling (BIM) based sustainability analysis for a construction project. *Proceedings of Sustainable Infrastructure Development and Management (SIDM)*, 5 April.

de Laat, R. and van Berlo, L. (2011). Integration of BIM and GIS: The development of the CityGML GeoBIM extension. In *Advances in 3D Geo-information Sciences*. Springer, Berlin, Heidelberg.

Meins, E., Wallbaum, H., Hardziewski, R., Feige, A. (2010). Sustainability and property valuation: A risk-based approach. *Building Research and Information*, 38(3), 280–300.

Monson, M. (2009). Valuation using hedonic pricing models valuation. *Cornell Real Estate Review*, 7, 62–73.

Mooya, M.M. (2016). *Estate Valuation Theory: A Critical Appraisal*. Springer-Verlag, Berlin, Heidelberg.

Morano, P., Rosato, P., Tajani, F., Manganelli, B., Di Liddo, F. (2019). Contextualized property market models vs. generalized mass appraisals: An innovative approach. *Sustainability*, 11(18), 4896.

Pagourtzi, E., Assimakopoulos, V., Hatzichristos, T., French, N. (2003). Real estate appraisal: A review of valuation methods, *Journal of Property Investment and Finance*, 21, 383–401.

Plume, J. and Mitchell, J. (2011). An urban information framework to support planning, decision-making and urban design. In *14th International Conference on Computer Aided Architectural Design – Designing Together*, Leclercq, P., Heylighen, A., Martin, G. (eds). CAADFutures, Liège.

RICS (2020). RICS Valuation–Global Standards. *Royal Institution of Chartered Surveyors (RICS)* [Online]. Available at: https://www.rics.org/globalassets/rics-website/media/upholding-professional-standards/sector-standards/valuation/rics-valuation--global-standards-jan.pdf.

Tauscher, H. (2020). Towards a generic mapping for IFC-CityGML data integration. *The International Archives of Photogrammetry, Remote Sensing and Spatial Information Sciences*, 44, 151–158.

TEGoVA (2016). European Valuation Standards (EVS) [Online]. Available at: https://tegova.org/european-valuation-standards-evs.

Tomić, H., Roić, M., Mastelić Ivić, S. (2012). Use of 3D cadastral data for real estate mass valuation in the urban areas. *3rd International Workshop on 3D Cadastres: Developments and Practices*, Shenzhen, China, 25–26 October.

Vitásek, S. and Zak, J. (2018). Cost estimating and building information modelling (BIM) in road construction. *Creative Construction Conference 2018*, 403–410. Budapest University of Technology and Economics.

Yu, S.M., Han, S.S., Chai, C.H. (2007). Modeling the value of view in high-rise apartments: A 3D GIS approach. *Environment and Planning B: Planning and Design*, 34(1), 139–153.

Zhang, J. (2019). Developing a comprehensive framework for property valuation using 3D and remote sensing techniques in China. Master's thesis, University of Twente, Enschede.

7

Semantic Segmentation of Airborne LiDAR Data for the Development of an Urban 3D Model

7.1. Introduction

In a context marked by urban sprawl, climate change, environmental degradation and resource depletion, cities are facing numerous issues. Monitoring urban dynamics, planning and management of urban spaces, environmental and climate studies are all issues that cities of the future must face in order to meet the needs of sustainable development.

3D modeling of urban spaces has become an increasingly common research and development topic. It is positioned as an essential method for the development of smart cities. The development of 3D digital models has the potential to meet the requirements of planning, monitoring and management of urban dynamics (Xiu *et al.* 2018). The urban 3D model allows not only the representation and 3D visualization of the urban space, but also a rich semantic modeling to accomplish several spatial and thematic analyses. 3D urban models appear as a potential solution that goes beyond the current limits of GIS models, which places them at the forefront of new needs in urban management.

The development of a 3D urban model requires a detailed 3D survey of the urban space. Airborne LiDAR is widely used for this purpose. It provides

Chapter written by Zouhair BALLOUCH and Rafika HAJJI.

geometric and radiometric information of objects in the form of point clouds. This acquisition system provides a large amount of accurate data with a very high level of detail, reliably and quickly. However, the transition from the point cloud to the digital model is tedious, time-consuming and essentially manual, and error-prone due to the large amount of data and methods that are difficult to automate. A major challenge today is to automate the processes leading to the construction of 3D urban models from point clouds, while reducing the costs associated with them. Also, the integration of semantic data obtained during the semantic segmentation phase of point clouds in a 3D model has the potential to meet the needs of urban space management.

This chapter is organized in five sections. In the second section, we highlight the workflow of elaboration of an urban 3D model from the semantic segmentation of airborne LiDAR data. The third section proposes a review of the literature on semantic segmentation methods of 3D point clouds by Deep Learning techniques. The fourth section is dedicated to the presentation of geometric and semantic modeling techniques for airborne LiDAR data, while section 5 proposes guidelines for our approach, followed by a conclusion.

7.2. From semantic segmentation to 3D modeling

To automatically build a 3D urban model from point clouds, two steps are essential. The first one involves automatically segmenting the point cloud to extract semantic classes. The second is the automatic modeling of the segmented semantic classes (Bellakaout 2016).

Semantic segmentation from point clouds is an essential step prior to the modeling of semantically rich 3D models. It consists of assigning a semantic label to each homogeneous group of point clouds or pixels in the case of images (Figure 7.1), in order to make the best use of the 3D data (Luo *et al.* 2020).

The semantic classes resulting from the semantic segmentation phase are then used in the 3D modeling process. From this point of view, the development of an automatic approach allowing the elaboration of a 3D urban model from the classified point clouds is a research question of great scientific and practical interest.

Figure 7.1. *Principle of semantic segmentation. For a color version of this figure, see www.iste.co.uk/hajji/BIM.zip*

The data acquired by the airborne LiDAR consists of an unstructured 3D point cloud that includes all the returns corresponding to urban objects (roads, buildings, etc.), and whose resolution and density vary according to the technical specifications of the used system. Before the raw data can be used, it must be pre-processed. This is essential to obtain coherent data; then, a semantic segmentation is necessary to subdivide the cloud of points into homogeneous subgroups and associate them with a semantic label, in order to integrate them into a 3D modeling process (Figure 7.2).

Unlike non-semantic segmentation methods (point clustering segmentation, primitive recognition segmentation), semantic segmentation techniques generate semantic information for each point (Xie *et al.* 2019), and are not limited to subdivision. Therefore, semantic segmentation is usually performed by supervised learning methods, including Machine

Learning-based segmentation (including Hierarchical Clustering, K-Means Clustering) and Deep Learning-based segmentation (including multi-view-based methods, voxel-based methods and direct methods that directly consume the point cloud) (Xie *et al.* 2019). Prior to the emergence of Deep Learning (DL) techniques, machine learning has been the only choice for semantic segmentation of 3D point clouds. Compared to machine learning, DL techniques have significant advantages: they are more efficient for processing large volumes of data and provide high accuracy results on public reference datasets.

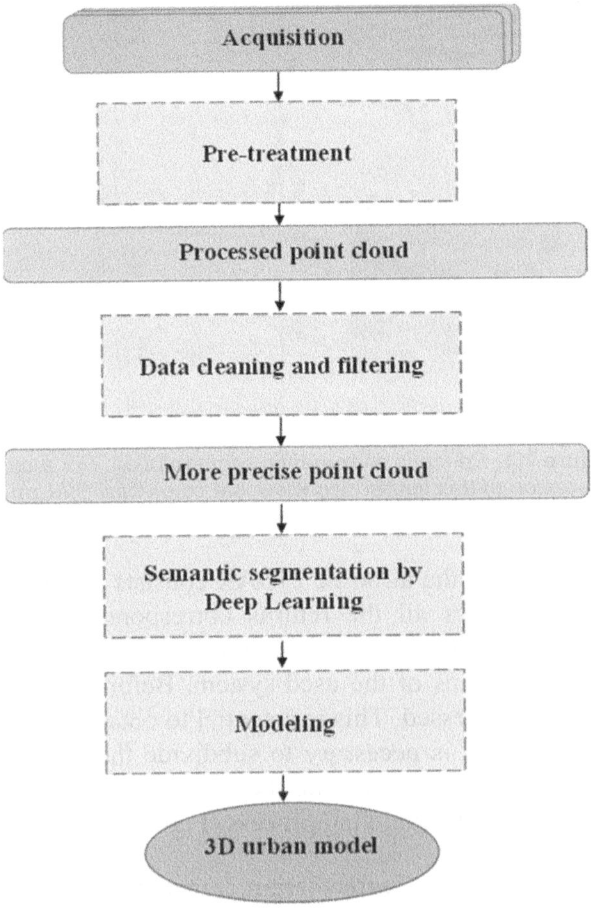

Figure 7.2. *The process of creating a 3D urban model from LiDAR point clouds. For a color version of this figure, see www.iste.co.uk/hajji/BIM.zip*

The results of the semantic segmentation of the point clouds into well-defined object classes (buildings, roads, vegetation, etc.) are used in the 3D modeling process.

Nowadays, 3D urban models are becoming richer and go far beyond simple geometric models whose main objective is 3D visualization. The introduction of semantic aspects of urban knowledge in the 3D model allows us to reach very high levels of semantic details. The 3D model must be the reference on which other data are grafted, which makes it possible to link a diversity of urban information (Billen 2013).

The main 3D modeling techniques encountered in the literature can be classified according to the used data (point cloud, drone images, etc.), the type of processing (mesh modeling, geometric modeling or hybrid) and the degree of human intervention (automatic, semi-automatic, etc.) (Bellakaout 2016).

There are two main 3D city models: design models and real world models. Design models are typically used in building construction and to meet the requirements of maximum detail in the architecture, engineering and construction industry. Real world models are geospatial information systems that represent the real world spatial objects and are widely represented in GIS applications (El-Mekawy 2010).

The scientific community is paying increasing attention to 3D city modeling approaches and their use for simulation and decision making purposes. Many 3D urban models such as OpenEarthView, OSM Buildings and ViziCities have been developed in recent years, and many studies have been conducted to evaluate them.

The Industry Foundation Classes (IFC) and City Geography Markup Language (CityGML) are object models built through object-oriented principles, with different types of classes and properties defined for the exchange of semantic information about urban objects. While IFC models buildings in terms of indoor elements, CityGML offers a rich modeling that integrates buildings and other urban objects (Kang and Hong 2018). The integration of the two semantic models, CityGML and IFC, for the modeling of urban models has the potential to answer many problems requiring indoor and outdoor modeling and management of relationships between urban objects, such as the case of energy simulations.

7.3. Semantic segmentation by Deep Learning

DL methods are revolutionizing the field of computer vision and are performing well in terms of accuracy and efficiency. Nowadays, semantic segmentation of large-scale acquired point clouds is one of the main problems in computer vision. With advances in DL methods, many algorithms have been developed in the literature to address the segmentation needs of LiDAR data (Castillo-Navarro *et al.* 2019)

7.3.1. *Semantic segmentation methods by DL*

In recent years, there has been a growing body of work aimed at adapting DL techniques for semantic segmentation of 3D point clouds. In this section, we review the main techniques developed in the literature, namely, SPGraph, PointNet, PointNet++, SnapNet, SegCloud and RandLA-Net.

SPGraph

SPGraph is a method that was proposed by Landrieu and Simonovsky (2018) for semantic segmentation of large point clouds simultaneously. The method is based on the principle of slicing the point cloud into simple shapes that are easy to segment. According to the results reported by Landrieu and Simonovsky (2018), this technique has shown a significant improvement in intersection over union compared to the literature, based on the tests performed on the two public datasets: “S3DIS” and “Semantic3D”.

PointNet

PointNet is a deep learning approach that processes the point cloud directly, without any resampling operations. Its efficiency, combined with its ease of implementation, grants it a perfect baseline for semantic segmentation of 3D point clouds. According to the results reported by Qi *et al.* (2016), the PointNet model is consistently accurate on an annotated dataset called “Stanford3D” compared to the state of the art.

PointNet++

PointNet++ is a hierarchical deep learning technique that allows the processing of a set of sampled points in a metric space. In order to test this approach, Qi *et al.* (2017) used four datasets: "ModelNet40", "MNIST", "SHREC15" and "ScanNet". The results show that the proposed approach is the most suitable for processing point clouds in a robust and efficient way compared to the existing methods (Qi *et al.* 2017).

SnapNet

SnapNet is an approach that relies on a two-dimensional sampling of the three-dimensional space of the point cloud through the generation of different views of the scene, and that offers the possibility to apply powerful 2D segmentation techniques, mainly fully convolutional networks. This methodology is composed of four main steps. The first one is a preprocessing step that consists of generating a cloud mesh; the second generates views; the third allows the production of 2D semantic segmentation maps, while the fourth consists of reprojecting the obtained semantic labels on the point cloud mesh (Boulch *et al.* 2018).

SegCloud

SegCloud is a method that offers the possibility to work more on cloud graphs via Conditional Random Fields (CRFs). Thus, it combines the advantages of neural networks, trilinear interpolation and fully connected conditional random fields to associate interpretations or semantics for subclouds of homogeneous points. This technique achieves equal or better performance than state-of-the-art methods based on neural networks, random forest algorithm and graphical models. The results show that CRFs provide fine-grained semantic information and show a great ability to extract object contours in a very accurate way (Tchapmi *et al.* 2017).

RandLA-Net

RandLA-Net is an approach to process large point clouds simultaneously (Hu *et al.* 2020). It shows good performance in terms of accuracy and computation time. This new deep learning methodology shows better performance compared to other semantic segmentation techniques based on tests performed on a set of datasets: "KITTI", "S3DIS" (which contains objects of building interiors) and "Semantic3D" (relative to outdoor objects).

The results show that this approach is up to 200x faster compared to the state of the art.

7.3.2. *Discussion*

In computer vision, semantic segmentation is defined as the assignment of an interpretation or semantics to each coherent region of an image or 3D point cloud. Many works using deep learning techniques have been developed for semantic segmentation of 3D point clouds and have shown great effectiveness (Qi *et al.* 2016; Qi *et al.* 2017; Landrieu and Simonovsky 2018; Qingyong *et al.* 2020).

The first attempts to use deep learning techniques for semantic segmentation of point clouds translated the algorithms used for semantic segmentation of images to the three-dimensional form; for example, the SnapNet method based on the conversion of LiDAR data into a series of two-dimensional images, which can generate a loss of information caused by the resampling operation. In the same sense, the SegCloud method organizes the point cloud in a structured voxel space to discard small details of the 3D shape. These regular point cloud representations are not able to describe the particular organization of the data with good accuracy, which limits the effectiveness of these semantic segmentation techniques (Landrieu and Simonovsky 2018).

Recently, several methods have been developed for the direct processing of 3D point clouds, but some of them are not applicable for large point clouds as in the case of PointNet and PointNet++ techniques. PointNet is a simple and robust method, but it is more suitable for the semantic segmentation of indoor data; it has, however, the disadvantage of the loss of contextual information. PointNet++ is an accurate method that has proven to be highly effective in many applications for classification and semantic segmentation of urban objects. Despite this, this method has some drawbacks, such as the requirement of a large memory.

Recent methods have been specifically designed for processing large numbers of points simultaneously, notably the SPG method. The advantage of this method is the fact that it slices the point cloud into simple shapes that are easy to classify, which facilitates the modeling of long distance interactions, and it takes into account the contextual information in the case of classification. However, the geometric partition and the construction of superpoint graphs are steps that require a significant amount of computation time. However, the RandLA-Net technique directly processes large point clouds simultaneously (one million points in a single run) with high performance (Cui *et al.* 2020).

Table 7.1 summarizes the advantages and disadvantages of the different semantic segmentation techniques based on DL presented above.

Technique	Advantages	Disadvantages
SnapNet	High efficiency	– Transformation of 3D data into 2D – Loss of information
SegCloud	High efficiency	Removal of small details from the 3D shape
PointNet	– Simplicity – Robustness	Loss of contextual information
PointNet++	High precision	– High cost – Inapplicability for large point clouds
SPG	Applicability for large point clouds	High-cost process of building superpoint graphs and geometric partitioning
RandLA-Net	High performance	High memory requirement

Table 7.1. *Advantages and disadvantages of semantic segmentation techniques of LiDAR point clouds based on DL*

7.4. Development of an urban 3D model

A 3D urban model is a geometric and semantic representation of an object or a set of urban objects (buildings, infrastructure, district, etc.). The 3D urban model makes it possible to integrate, organize and exchange data between different stakeholders for an efficient management of cities. The interdisciplinarity and interoperability of the used data make this urban model a tool for collaborative design, simulation, analysis, multitemporal management and decision-making.

The production of 3D urban models has developed remarkably in recent years. Their richness and degree of accuracy depends on the mode of acquisition of geometric data and the adopted process for segmentation and modeling.

7.4.1. *Geometric modeling from the 3D point cloud*

The goal of 3D modeling is to create and build links between cloud feature points to realize an object or a model (Boulaassal 2010). To model an object from LiDAR data, three main approaches exist in the literature: geometric modeling, mesh modeling and hybrid modeling. In addition to these three approaches, we cite the parametric modeling which characterizes BIM models.

Geometric modeling consists of identifying the adequate model from a set of predefined models, where each model is described by a certain number of parameters, and each object is approximated by a geometric shape, such as a cylinder or cone. The main advantage of geometric modeling is to provide geometric models without visible deformations with a limited computation time. However, the major disadvantage of this modeling lies in the fact that it only offers a reduced number of element models and is limited to elements with simple geometric shapes. Moreover, these models are only reliable if the density of points is relatively high (more than 2 points/m^2) (Tarsha Kurdi 2008). The modeling of complex surfaces and architectural details of buildings, such as ornaments or moldings, can only be done with mesh models that allow for a more accurate representation of the objects' shapes.

Mesh modeling is based on 3D point clouds without referring to other sources of information or predefined models, using, in most cases, triangulation algorithms, mainly Delaunay triangulation (Reboul 2014). This type of modeling is adopted for the representation of complex urban surfaces, such as historical monuments with a very high level of detail. The advantage of this modeling lies in its fidelity to reality, by the fact of modeling a surface, whatever its geometric form, contrary to geometric modeling which requires the approximation of surfaces by predefined shapes. However, mesh modeling requires a large memory for storing the generated shapes.

Hybrid modeling is a compromise between geometric and mesh modeling. In this method, the reconstructed surface combines both structured parts describing geometric primitives and freeform parts representing objects that cannot be described by geometric modeling.

The parametric modeling adopted by BIM represents objects according to predefined rules and parameters that will determine the geometry, as well as the relationships between elements (Eastman *et al.* 2011). It allows us to design more formal architectures and to produce an infinity of geometrical forms that facilitates the projection of complex shapes. Parametric modeling offers possibilities that go far beyond geometric models and the simple spatialization of relationships between objects. The BIM model can also provide information on a given project (costs, energy qualities, materials, etc.). Many scientific works using LiDAR point clouds for BIM modeling have been proposed in the literature and have shown very promising results (Xue *et al.* 2019; Yang *et al.* 2020; Romero-Jarén and Arranz 2021).

7.4.2. Semantic modeling

Several standards such as CityGML and IFC have been developed for the deployment of a 3D urban model. CityGML is an international standard for 3D urban modeling that was validated in 2008 by the Open GIS Consortium (OGC). CityGML defines a model for representing the relationships between the main objects of the urban space under geometric, topological and semantic aspects. Thanks to its semantic richness and modularity, it offers a great flexibility of representation. The adoption of the CityGML standard to elaborate a 3D urban model has contributed to the resolution of the interoperability problem in information exchange. The objective of CityGML is mainly to contribute to a common understanding and sharing of data from different 3D models. CityGML proposes, in its "Building module", five levels of detail (LoD0, LoD1, LoD2, LoD3 and LoD4) that vary from the simple footprint to the complete modeling of the building interior (Figure 7.3), where geometric and semantic details increase at each level.

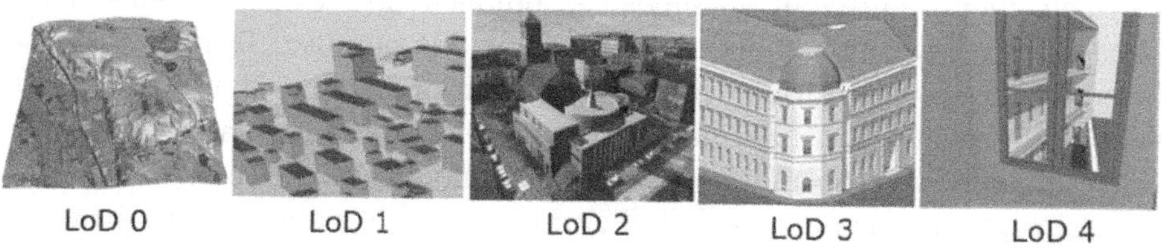

Figure 7.3. *The five levels of detail (LOD) defined by CityGML (OGC, 2012). For a color version of this figure, see www.iste.co.uk/hajji/BIM.zip*

IFC is an object-oriented modeling format that was developed by buildingSMART. It defines a set of object classes (about 900 classes) to model information about the objects and elements of a BIM model. It is capable of representing information about the geometry of each object, in addition to semantic information such as colors, materials, mechanical and physical properties, textures and prices (Mazars 2017), and also relationships with other objects. The schema provides a framework that supports information sharing across different applications and disciplines. The architecture of the IFC schema has been designed to have four different hierarchical levels; each level has a conceptual layer in the form of an "individual schema" defining the modularity of the structure (El-Mekawy 2010). The model must be flexible and dynamically updated to allow compatibility between different versions of the standard.

IFC and CityGML are two examples of the most developed and established semantic models for interior (building elements) and exterior (urban objects), respectively, which can address many issues that require modeling of the interior and exterior, including urban planning studies (Thompson *et al.* 2011), energy management (Wu *et al.* 2014), water and sewage management (Hijazi *et al.* 2011) and indoor and outdoor navigation (Hagedorn *et al.* 2009). The integration of IFC and CityGML building models presents some difficulties that may arise when translating information from one to the other. Professionals from both have made significant attempts to integrate CityGML and IFC models to produce useful common applications. Most of these attempts, however, use a one-way method (mainly from IFC to CityGML) for the conversion process (El-Mekawy 2010). Different research and projects have shown that the two models can be integrated and semantic information can be exchanged between them. Different methods, techniques and even software have already been investigated and developed for this purpose. The use of these formats allows structuring, prioritizing and organizing the information, taking into account the different classes, attributes and relationships, in order to facilitate the interoperability of 3D urban models in all the different BIM software on the market, so that they are made available to the different stakeholders.

7.4.3. *Discussion*

Modeling techniques can be grouped into three categories: geometric modeling, mesh modeling and hybrid modeling. Geometric modeling techniques allow for the representation of objects using simple geometric primitives (planes, cylinders, straight lines, etc.), and therefore make it possible to reduce the volume of the generated data, as well as to embed semantic data in the model. However, the result of geometric modeling depends on the parameterization of the algorithms used in each step. Also, this representation remains coarse, neglecting the fine details of the surface. Mesh modeling has the main advantage of highlighting the fine details of the digitized surface. However, the data remains voluminous, which makes the interpretation and manipulation of the model laborious for the user. Another limitation of this type of modeling is the impossibility of embedding semantic data on the model because no shape or feature recognition is performed.

In view of the limitations of 3D modeling methods reported in the literature, we argue that the development of 3D urban models is still an active research trend that aims to achieve robust and efficient methods to implement geometrically and semantically rich 3D urban models. Our research attempts to address this challenge by developing a new approach to semantic segmentation of 3D point clouds, in order to build 3D urban models with a high level of geometric and semantic detail.

7.4.4. *Our approach*

Today, the production of 3D urban models is booming. However, the developed digital models do not yet integrate sufficiently rich semantics in terms of urban knowledge.

To meet this need, the exploitation of the results of semantic segmentation has potential for the implementation of 3D urban models that make it possible to model and represent semantic data in an urban 3D space, and thus offer an effective tool for consultation, instruction and urban planning to the developers and decision-makers.

The construction of urban models from LiDAR point clouds requires a robust processing chain to extract the maximum semantic information from the point cloud, with high accuracy and good algorithmic performance. In our research, we propose to design a methodology integrating the results of semantic segmentation of airborne LiDAR data for 3D modeling of urban space. Figure 7.4 summarizes the steps of our approach.

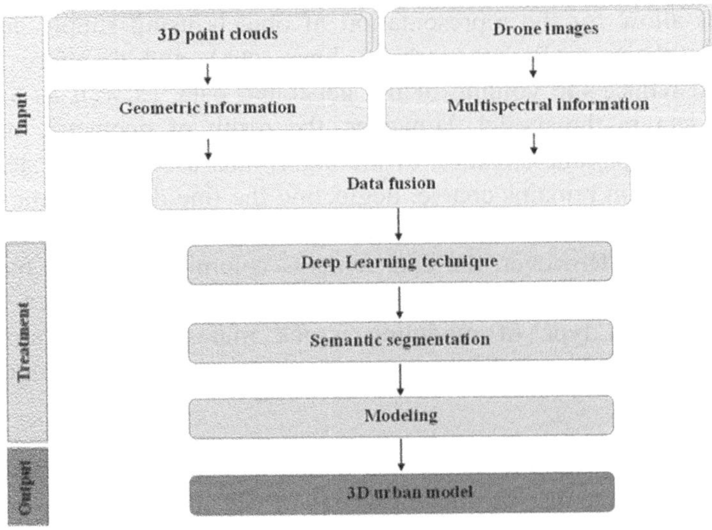

Figure 7.4. *General methodology of our approach. For a color version of this figure, see www.iste.co.uk/hajji/BIM.zip*

In the proposed approach, we focus on the development of a new method for semantic segmentation of airborne LiDAR data based on DL. The remarkable emergence of DL methods shows a good performance in semantic segmentation of 3D point clouds. The main objective of this research is to reach high accuracy and good performance in the semantic segmentation process.

Several research teams have stated that the combination of 3D point clouds with other sources (drone images, satellite images, etc.) is promising (Gilani *et al.* 2015; Griffiths and Boehm 2019), thanks to the planimetric continuity of the images and the altimetric accuracy of 3D point clouds. This motivates us to carry out our research in this direction, where we propose to design an automatic approach of semantic segmentation through a hybrid method that integrates point clouds and images for the automatic recognition of objects in an urban scene integrating DL (Ballouch *et al.* 2020). It is based on the use of raw point clouds without any resampling operation, in order to preserve the topological relationships and the original characteristics of the cloud. The projected objective is to merge the geometric information from the point clouds and the multi-spectral information from the UAV images, which will help in the differentiation of the different objects present in the urban scene.

Through this research, we want to adopt a DL-based methodology that will be able to extract all of the 3D objects present in a scene with high geometric accuracy and good semantic richness. The obtained results will be integrated in a modeling process exploiting the semantic classes resulting from the segmentation. A set of tests related to different urban contexts will be performed to validate our approach.

7.5. Conclusion

Several techniques for semantic segmentation and modeling of 3D point clouds acquired in urban environments have been developed in the literature. In this chapter, we have analyzed a set of scientific works using deep learning techniques that have been developed for the semantic segmentation of 3D point clouds. In addition, we have discussed the limitations and performances of the main 3D geometric modeling techniques.

The integration of urban knowledge in 3D city models has the potential to meet the needs of the sustainable development of cities. To meet the need for the development of automatic and robust approaches for the semantization of large-scale digital models, we propose to design and develop a Deep Learning-based approach for the semantic segmentation of data from an airborne LiDAR. Our results can then be integrated into a modeling process for the semantic enrichment of 3D geometric models.

7.6. References

Ballouch, Z., Hajji, R., Ettarid, M. (2020). The contribution of Deep Learning to the semantic segmentation of 3D point-clouds in urban areas. *2020 IEEE International conference of Moroccan Geomatics (Morgeo)*, 1–6.

Bellakaout, A. (2016). Extraction automatique des bâtiments, végétation et voirie à partir des données LiDAR 3D. PhD Thesis, Institut Agronomique et Vétérinaire Hassan II, Morocco.

Billen, R. (2013). L'avenir des modèles 3D urbains. *GeoPlatform*, 22–25.

Boulaassal, H. (2010). Segmentation et modélisation géométrique de façades de bâtiments à partir de relevés laser terrestres. PhD Thesis, Université de Strasbourg.

Boulch, A., Guerry, J., Saux, B.L., Audebert, N. (2018). Snapnet: 3D point cloud semantic labeling with 2D deep segmentation networks. *Computers Graphics*, 71, 189–198.

Castillo-Navarro, J., Le Saux, B., Boulch, A., Lefèvre, S. (2019). Réseaux de neurones semi-supervisés pour la segmentation sémantique en télédétection. *Colloque GRETSI sur le Traitement du Signal et des Images*, Lille, France [Online]. Available at: https://hal.archives-ouvertes.fr/hal-02343961/document.

Cui, Y., Chen, R., Chu, W., Chen, L., Tian, D., Li, Y., Cao, D. (2020). Deep learning for image and point cloud fusion in autonomous driving: A review. *IEEE Transactions on Intelligent Transportation Systems*. arXiv:2004.05224.

Eastman, C., Teicholz, P., Sacks, R., Liston, K. (2011). *BIM Handbook: A Guide to Building Information Modeling for Owners, Managers, Designers, Engineers and Contractors*, 2nd edition. John Wiley & Sons Inc, Hoboken, NJ.

El-Mekawy, M. (2010). Integrating BIM and GIS for 3D city modeling: The case of IFC and CityGML. *TRITA SoM 2010-11*.

Gilani, S.A.N., Awrangjeb, M., Lu, G. (2015). Fusion of LiDAR data and multispectral imagery for effective building detection based on graph and connected component analysis. *The International Archives of the Photogrammetry, Remote Sensing and Spatial Information Sciences*, Volume XL-3/W2.

Griffiths, D. and Boehm, J. (2019). Improving public data for building segmentation from Convolutional Neural Networks (CNNs) for fused airborne LiDAR and image data using active contours. *ISPRS Journal of Photogrammetry and Remote Sensing*, 154(2019), 70–83.

Hagedorn, B., Trapp, M., Glander, T., Döllner, J. (2009). Towards an indoor level-of-detail model for route visualization. *10th International Conference Mobile Data Management: Systems, Services and Middleware*. IEEE, Taipei, 692–697.

Hijazi, I., Ehlers, M., Zlatanova, S., Becker, T., van Berlo, L. (2011). Initial investigations for modeling interior utilities within 3D geo context: Transforming IFC-interior utility to CityGML/Utility Network ADE. In *Advances in 3D Geo-information Sciences*. Kolbe, T.H., König, G., Nagel, C. (eds). Springer, Berlin, Heidelberg.

Hu, Q., Yang, B., Xie, L., Rosa, S., Guo, Y., Wang, Z., Trigoni, N., Markham, A. (2020). RandLA-Net: Efficient semantic segmentation of large-scale point clouds. *Proceedings of the IEEE/CVF Conference on Computer Vision and Pattern Recognition*, 11108–11117.

Kang, T.W. and Hong, C.H. (2017). IFC-CityGML LOD mapping automation using multiprocessing-based screen-buffer scanning including mapping rule. *KSCE Journal of Civil Engineering*, 22(2), 373–383.

Landrieu, L. and Simonovsky, M. (2018). Large-scale point cloud semantic segmentation with Superpoint graphs. *The IEEE Conference on Computer Vision and Pattern Recognition (CVPR)*, 4558–4567.

Luo, H., Khoshelham, K., Fang, L., Chen, C. (2020). Unsupervised scene adaptation for semantic segmentation of urban mobile laser scanning point clouds. *ISPRS Journal of Photogrammetry and Remote Sensing*, 169, 253–267.

Mazars, T. (2017). Exploitation du BIM pour la modélisation Chronographique de la planification et la simulation 4D. Thesis, Ecole de technologie supérieure, Université du Québec.

Qi, C.R., Su, H., Mo, K., Guibas, L.J. (2016). Pointnet: Deep learning on point sets for 3d classification and segmentation. *CoRR*, abs/1612.00593.

Qi, C.R., Yi, L., Su, H., Guibas, L.J. (2017). PointNet++: Deep hierarchical feature learning on point sets in a metric space. arXiv:1706.02413v1 [cs.CV].

Reboul, R. (2014). Segmentation, classification et modélisation à partir de données issues de capteurs dynamiques. Master’s thesis, HESAM University.

Romero-Jarén, R. and Arranz, J.J. (2021). Automatic segmentation and classification of BIM elements from point clouds. *Automation in Construction*, 124, 103576.

Tarsha Kurdi, F. (2008). Extraction et reconstruction de bâtiments en 3D à partir de relevés LiDAR aéroportés. PhD Thesis, Université Louis Pasteur Strasbourg I.

Tchapmi, L.P., Choy, C.B., Armeni, I., Gwak, J., Savarese, S. (2017). Segcloud: Semantic segmentation of 3D point clouds. *2017 International Conference on 3D Vision (3DV)*, 537–547.

Thompson, E.M., Horne, M., Lockley, S., Cerny, M. (2011). Towards an information rich 3D city model: Virtual Newcastle Gateshead GIS Integration. *CUPUM ‘11 12th International Conference on Computers in Urban Planning and Urban Management*, Alberta, Canada.

Wu, W., Yang, X., Fan, Q. (2014). GIS-BIM based virtual facility energy assessment (VFEA) Framework development and use case of California State University, Fresno. *Proceedings of the 2014 International Conference on Computing in Civil and Building Engineering*, Orlando, FL.

Xie, Y., Tian, J., Zhu, X.X. (2019). Linking points with labels in 3D: A review of point cloud semantic segmentation. *IEEE Geoscience and Remote Sensing Magazine*, 8(4), 38–59.

Xiu, H., Vinayaraj, P., Kim, K.-S., Nakamura, R., Yan, W. (2018). 3D semantic segmentation for high-resolution aerial survey derived point clouds using deep learning (Demonstration). In *Information Systems (SIGSPATIAL'18)*, Farnoush, B.-K., Erik, H. (eds). ACM, New York.

Xue, F., Lu, W., Chen, K., Webster, C.J. (2019). BIM reconstruction from 3D point clouds: A semantic registration approach based on multimodal optimization and architectural design knowledge. *Advanced Engineering Informatics*, 42, 100965.

Yang, L., Cheng, J.C., Wang, Q. (2020). Semi-automated generation of parametric BIM for steel structures based on terrestrial laser scanning data. *Automation in Construction*, 112, 103037.

8

BIM for the Renovation of Urban Spaces

8.1. Introduction

Nowadays, cities are witnessing urban expansion at an unprecedented scale, as well as the scarcity of land resources. In response to such urban pressure, there is a great tendency towards the use of space in three dimensions, both in height (vertical constructions) and in depth (e.g. underground networks). In addition to the construction of buildings, a large number of urban infrastructure projects are multiplying day by day. These infrastructures are often poorly documented and dispersed in different sectors and areas, which confronts managers and decision-makers with problems of efficient management of the city. These problems cannot be solved solely based on two-dimensional plans and data, but require rich 3D urban models for their modeling and management (Xu *et al.* 2014).

Due to lack of maintenance and/or poor planning, urban spaces deteriorate and lose their attractiveness. They become the subject of urban renewal programs, which must take into consideration all of the components of this space, in order to promote a global and integrated improvement. A successful urban renewal is based on a good knowledge of the site in question, which enables appropriate actions to be taken in order to meet the actual needs of the site and its occupants. It also allows the management and monitoring of this site, to ensure its maintenance after completion of the work. Hence, the importance of the 3D urban digital model as a tool for modeling the existing space, and facilitating new designs and management of the modeled structures.

Chapter written by Oumayma MOUFID.

Certainly, the field of urban planning is currently moving towards 3D modeling of buildings, notably through BIM, as well as 3D modeling of a city's infrastructure and equipment located on the Earth's surface. Nevertheless, a true urban model implies the modeling of any object above or below the ground, including underground utilities (drinking water, storm water, sewage, natural gas, electricity, etc.), which constitute an essential infrastructure for all cities. However, many components are aging, in unfamiliar locations with congested configurations and under unknown conditions, resulting in numerous degradations (Huston *et al.* 2017).

Due to the great contribution of 3D urban modeling, various standards are being developed to enable the representation, storage and exchange of data related to the components of the city, including CityGML, IFC, LandInfra and so on, and several institutions are continuously developing data model specifications to support certain objects in the urban space, as needed. However, no single standard supports the entire 3D modeling urban space.

8.2. Urban space: problems and dysfunctions

An urban space is composed of several elements connected above and below the ground. These are buildings, facilities, roads, public spaces, infrastructure, networks and installations and so on. The way in which these structural elements are distributed, managed and used makes the difference between an equipped planned space and an under-equipped spontaneous space.

In addition to the socio-economic fragility of an urban space that primarily affects humans, and causes all kinds of deficits and marginalization, there is a spatial and urban fragility that is manifested by a range of problems and malfunctions, causing the degradation of one or more components of the urban space. This includes:

The built environment

The degradation of the built environment is a major issue for any urban space, not only because it affects the urban and architectural image of the concerned site, but also because it puts human lives at risk.

A healthy building is the one that is built according to standards and is maintained periodically. In the absence of these two elements, there are two urban problems: "illegal housing" and "housing in danger of collapse". The first encompasses housing outside the norms of construction and without respect for the applicable urban regulations, that is, illegal housing and shanty towns. The second concerns the buildings threatening ruin, which no longer meet the necessary guarantees of solidity, due to the dilapidation of one of their main components or external environment. This phenomenon affects all types of materials, but in particular, old and historical materials, and unfortunately, in the absence of maintenance, upkeep and archiving, several spaces and historical sites lose their heritage and architectural value, and thus their identity.

Road infrastructure

Mobility is a major issue for any urban space. Without proper design and planning, several issues can devastate.

An under-equipped area in terms of road networks mainly generates traffic and parking problems. In addition, excessive automobile congestion caused by heavy traffic (through traffic and feeder traffic) reflects multiple problems, namely: air pollution and adverse environmental effects, degradation of accessibility, quality of public transport service weakened by poor traffic conditions, insufficient, anarchic and poorly organized parking supply, road insecurity for pedestrians, poorly exploited public spaces and so on.

Public spaces

Public spaces – that is, public squares, green spaces and so on – are essential components for each planned and equipped urban space. A public space is considered fragile for one of several reasons if its degradation is not attended to at the time.

As its name indicates, a public space is intended for the public. Nevertheless, many factors do not favor its appropriation by all users, including the poor choice of location, its random architecture that deteriorates the urban landscape and flaws related to management, including conflicts of use, squatting by non-regulatory trade, as well as disparate street furniture and signage.

Facilities and services

Facilities (cinema, leisure center, etc.) and services (shops and activities) are a source of attractiveness for each urban area. An area that is not served by local facilities and services risks devitalization, a decline in its population and aging.

Moreover, the lack of facilities that weaken an urban space, but even if they exist, the poor choice of location, the degraded state of the built environment and the disorder in management also accentuate the phenomenon of urban fragility.

Miscellaneous facilities and networks

Malfunctions in the various networks (sanitation, drinking water, electricity, etc.) are problems in themselves and cause others. The difficulty in the management of these installations arises because they are underground infrastructures whose surveyal, localization and management are not as obvious as when they are above ground.

8.3. Urban renewal approaches

In the field of building and urban planning, there are different approaches to interventions. International charters and organizations responsible for planning and urban intervention operations use a variety of terminology to describe any type of approach. One speaks in particular of rehabilitation, conservation-restoration, requalification, restructuring, renovation and so on.

When acting only on the built environment, two approaches are generally found:

– Rehabilitation: an action that targets the functional and aesthetic enhancement of a building by restoring its original use value and architectural characteristics that it has lost, either because it has been abandoned or because it has been used for purposes not adapted to its initial vocation.

– Conservation-restoration: two terms that are always attached to the built heritage. They describe the set of measures and actions whose objective is to safeguard the tangible cultural heritage, while guaranteeing its accessibility to present and future generations. Conservation includes anticipatory conservation, which involves measures that avoid and minimize future deterioration or loss, and curative conservation, which consists of actions aimed at stopping an active process of deterioration of a building or its reinforcement.

When it comes to interventions affecting the entire urban space, we are faced with a set of approaches including:

– Requalification: an operation that consists of modifying the physical qualities of a space in order to give it a new vocation. Generally, it is accompanied by the consolidation of the concerned space, the reconversion of its equipment and the reallocation of its public places, as well as other actions concerning the choices retained in the new conception.

– Restructuring: an approach that consists of the in-depth redesign of a space in order to achieve better urban performance. It is generally accompanied by the implementation and/or improvement of the state of the basic infrastructure (roads, various networks) and the reconstruction of housing.

– Renovation: this is the most integrated approach, aiming to enhance the value of an urban area by acting on all of its components. This is achieved through rehabilitation and restoration operations, infrastructure development, the provision of basic facilities and the construction of new housing, as well as land development operations, while taking into consideration the respect of environmental protection conditions.

The success of these interventions requires a combination of both human (architects, engineers, etc.) and technical skills, including efficient and effective tools and processes. Today, geospatial data management tools are developing, and the urban planning world must take advantage of them.

8.4. BIM/CIM for urban renewal

It is useful to use the BIM/CIM model to draw possible renovation scenarios by targeting all elements constituting an urban space, in order to improve and enhance it.

8.4.1. *Methodology*

The objective of this study is to develop a 3D digital model of an urban space subject to urban renewal, with all of its components above and below ground. Thus, the model is used to diagnose the space in question and extract its state of affairs on all levels (built environment, mobility and transport, facilities, etc.), as well as plan and manage the projects and actions to be undertaken in this space.

To achieve this, the scientific approach (Figure 8.1) must begin with the design of a spatial and integrated data model that supports and enables the 3D modeling of the urban space as a whole. Then comes the elaboration of the digital urban model using 3D survey tools (3D scan, drone, etc.), which will be an entry point for the implementation of the proposed data model. This model will then be the object of simulations and analysis of urban renewal scenarios for the concerned area.

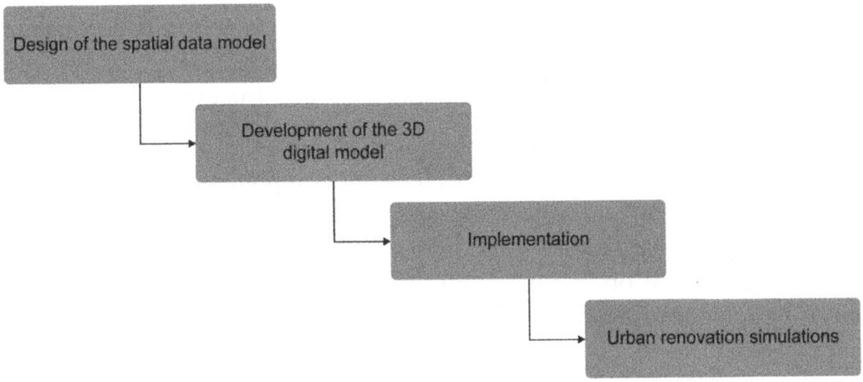

Figure 8.1. *Diagram of the methodological approach*

The data model is designed for the digital modeling of a space with all of its urbanistic components. Models already exist, some of which are mentioned in the previous section, but each one represents a few components. Thus, we must proceed by integrating and merging information from these models into a common model. The concept of this model is to:

– Integrate the IFC and CityGML standards in order to capture the large amount of data related to buildings from the BIM model via the IFC standard, and to have the data of other elements of the urban space (roads, green spaces, urban furniture, etc.), while including the GIS model.

– Design a model capable of modeling underground infrastructures in 3D (various networks, installations, tunnels, geological and hydrological elements, etc.) by integrating existing standards and specifications and adding the necessary elements.

The 3D modeling is based on the surveying of the existing objects. The latter, when it is about visible and accessible elements, is mainly based on the following methods:

– Direct methods based on photogrammetry (aerial by plane or by drone, or methods based on terrestrial, aerial or Mobile Mapping Systems (MMS)).

– Hybrid methods which involve the integration of data from both methods.

The choice of the method, equipment configuration and surveying parameters (image resolution, point cloud density, etc.) depends on the desired accuracy and the level of detail envisaged in the digital model.

When it comes to the study of underground structures, such as sewerage networks and electricity, other techniques are used, including modeling based on 3D plans or surveying by magnetic or electronic sensors (e.g. GeoRadar).

The collected data (images, point clouds, etc.) undergo georeferencing, cleaning and consolidation. Then, each urban component is modeled separately. The buildings are modeled in the form of a BIM model with the integration of all related semantic data. Finally, the urban objects of interest for the urban renewal will be modeled within a rich 3D model of the urban space.

After consolidation of the model, the data model designed in the previous phase is implemented and each entity is linked to its physical element in the model.

The implementation will be done in a 3D GIS environment, allowing the management and interrogation of both geometric and semantic data. Thus, it will facilitate the planning of urban renewal operations at the site level.

8.5. Renovation process

The renovation of an urban space goes through three main phases that call upon the digital urban model thus developed.

"Diagnostic" phase

The objective of this first phase is to have an inventory of the concerned space. It is thus necessary to identify the various problems and dysfunctions from which the site suffers, as well as its assets and its interactions with its environment.

The digital model, allows the identification of existing anomalies in the built environment (deterioration, deformation, lack of facilities, etc.), including problems related to mobility and road traffic, and those concerning the channeling and installation of various networks (sanitation, drinking water, etc.). It also allows the detection of the lack of facilities and public spaces, and the land (or buildings to be reconverted) that can accommodate them, while taking into account the land data. Moreover, the 3D model represents the most efficient tool for archiving the urban and architectural aspects of buildings with heritage value.

"Planning" phase

The planning stipulates the declension of a vision of renovation in actions and integrated projects, aiming for the valorization of the space in question. After identifying the assets and anomalies, it is necessary to design the set of projects to be injected into this space.

In this phase, all projects are modeled in 3D within the digital model. The latter facilitates the design because it offers all of the data to be taken into consideration concerning the site and its environment. It will thus allow, a better design that covers all levels.

The result is an "Urban Renovation Model"; a virtual 3D digital model, which draws the new design of the site.

"Execution and management" phase

This phase is about implementing the designed actions and projects. After the execution of all the projects, a re-modeling is carried out, if necessary, in case of modifications applied on some parts which differ from what was planned in the Urban Renovation Model. This way, the space in question will always have a digital model to update and interrogate if necessary, which will facilitate its management and maintenance.

8.6. Conclusion

This study, whose objective is to show the great importance of 3D digital modeling in the renovation of urban spaces, proposes a conceptual framework based on BIM and CIM. An approach that aims to integrate all the elements of an urban space, whether above or below ground, in a common model, thus establishing a digital urban model that models the whole concerned site.

The data model designed is based on the integration of a set of standards and specifications, the main one being CityGML. It gathers a large part of the urban and spatial elements and also allows the integration of the elaborated model in a GIS environment that will facilitate the management, and will make it possible to produce different simulations. The use of the IFC standard is also essential in the designed data model, because of the great richness it offers in terms of data related to buildings and facilities.

The approach developed will thus provide urban planning organizations and stakeholders in the field of building and urbanism, with a new way to renovate weakened urban spaces and a digital 3D database to identify anomalies, thus conceiving an improved and valorized space.

8.7. References

Huston, R.D., Zhang, Y., Xia, T. (2017). Urban underground infrastructure mapping and assessment. *Conference Paper in the Event: SPIE Smart Structure and Materials + Non-destructive Evaluation and Health Monitoring*, Portland, OR.

Xu, X., Ding, L., Luo, H., Ma, L. (2014). From building information modelling to city information modeling. *Journal of Information Technology in Construction*, 19(17), 292–307. ISSN 1874–4753.

Conclusion

BIM (Building Information Modeling) represents a major innovation in construction engineering that is based on a new business paradigm leading to major changes in the way buildings are designed, constructed, managed and operated. BIM has environmental, economic and social components that must be integrated into a larger space, putting the building in its geographical and urban context. This is how the concept of multi-scale BIM emerges to integrate different spatial and thematic dimensions in the intelligent and sustainable management of the city.

Multi-scale BIM must capitalize on a very rich urban knowledge by taking advantage of data models developed for BIM and 3D GIS (Geographic Information Systems). There are two perspectives when it comes to development. The first relates to 3D acquisition, modeling and semantic enrichment to produce the 3D base of multi-scale BIM, where different dimensions and scales of analysis are integrated. This requires the development of processing, storage and visualization methods adapted to the massive and heterogeneous data of the urban space. The other perspective focuses on the integration of multi-source geometric and thematic data, often distributed among several actors, in different formats and models for the geometric construction and semantic enrichment of digital models.

The challenge is even greater when it comes to interoperability, format translation and the development of formalized processes for exchanging data and practices. This need merits the development of data production approaches and acquisition protocols that meet common specifications to

---

Conclusion written by Rafika HAJJI and Hassane JARAR OULIDI.

minimize problems related to the exchange of information. The "Level of Information Needed" is a new concept to standardize the data exchange of BIM models in a standardized and need-based framework.

Territorial intelligence is nowadays a very fashionable research field thanks to the increased awareness of the urgency to react to sustainable development and smart cities. Given the complexity and the large number of parameters that interact in urban issues, this field of research is multidisciplinary and thus calls upon several skills. Indeed, this field is invested in by urbanists, planners, geographers, architects, surveyors, geomaticians, climatologists and so on. Each of them brings a different scale of understanding to this field. Each of them brings a scale of analysis to the problem of sustainable cities. As a result, levels of expertise in BIM and 3D GIS are necessary among the different actors in order to develop an ecosystem that values efficient and effective sharing of territorial knowledge, as well as the co-development of processes, practices, standardized and open applications.

Multi-scale BIM can benefit from developments and expertise in the world of architecture and engineering and of BIM, or from those of the world of GIS. It must take into account the standardization work in progress, which is moving towards bringing the two worlds closer together through the enrichment of the two standards IFC and CityGML. Finally, it is the role of city specialists and managers to define the bases of territorial management to which the multi-scale BIM must respond.

The development of multi-scale BIM should take advantage of the established knowledge of BIM (Chapter 1), of the great revolution in 3D acquisition techniques (Chapter 2), in 3D modeling techniques for the development of digital mock-ups (Chapter 3) and of recent developments in open BIM standards (Chapter 4), as well as of attempts to converge the worlds of BIM and GIS under the concept of GeoBIM (Chapter 5). However, the challenge lies in the capitalization of knowledge and the development of urban simulation solutions capable of meeting the challenges of smart cities.

Today, multi-scale BIM must be integrated into a rich ecosystem integrating several technologies such as Big Data, artificial intelligence, the Internet of Things and digital twins. All this must contribute to the development of intelligent, sustainable and connected urban spaces.

In this book, we have tried to cover as broadly as possible the theoretical and practical concepts around the development of multi-scale BIM and thus establish the basis for their implementation for intelligent territorial management. We have illustrated, through practical case examples, some contexts and practical aspects of their implementation. Other aspects, related to the simulations, can be further explored.

List of Authors

Rafika HAJJI
Geomatic Sciences and
Topographic Engineering
IAV Hassan II
Rabat
Morocco

Hassane JARAR OULIDI
Geomatic Sciences
EHTP
Casablanca
Morocco

Zouhair BALLOUCH
IAV Hasssan II
Rabat
Morocco

Siham EL YAMANI
IAV Hasssan II
Rabat
Morocco

Oumayma MOUFID
EHTP
Casablanca
Morocco

Index

L, M

N, O

P, R

S, T

U, V

Other titles from

iSTE

in

Systems and Industrial Engineering – Robotics

2020

BRON Jean-Yves
System Requirements Engineering

KRYSINSKI TOMASZ, MALBURET FRANÇOIS
Energy and Motorization in the Automotive and Aeronautics Industries

PRINTZ Jacques
System Architecture and Complexity: Contribution of Systems of Systems to Systems Thinking

2019

ANDRÉ Jean-Claude
Industry 4.0: Paradoxes and Conflicts

BENSALAH Mounir, ELOUADI Abdelmajid, MHARZI Hassan
Railway Information Modeling RIM: The Track to Rail Modernization

BLUA Philippe, YALAOU Farouk, AMODEO Lionel, DE BLOCK Michaël, LAPLANCHE David
Hospital Logistics and e-Management: Digital Transition and Revolution

BRIFFAUT Jean-Pierre
From Complexity in the Natural Sciences to Complexity in Operations Management Systems
(Systems of Systems Complexity Set – Volume 1)

BUDINGER Marc, HAZYUK Ion, COÏC Clément
Multi-Physics Modeling of Technological Systems

FLAUS Jean-Marie
Cybersecurity of Industrial Systems

JAULIN Luc
Mobile Robotics – Second Edition Revised and Updated

KUMAR Kaushik, DAVIM Paulo J.
Optimization for Engineering Problems

TRIGEASSOU Jean-Claude, MAAMRI Nezha
Analysis, Modeling and Stability of Fractional Order Differential Systems 1: The Infinite State Approach
Analysis, Modeling and Stability of Fractional Order Differential Systems 2: The Infinite State Approach

VANDERHAEGEN Frédéric, MAAOUI Choubeila, SALLAK Mohamed, BERDJAG Denis
Automation Challenges of Socio-technical Systems

2018

BERRAH Lamia, CLIVILLÉ Vincent, FOULLOY Laurent
Industrial Objectives and Industrial Performance: Concepts and Fuzzy Handling

GONZALEZ-FELIU Jesus
Sustainable Urban Logistics: Planning and Evaluation

GROUS Ammar
Applied Mechanical Design

LEROY Alain
Production Availability and Reliability: Use in the Oil and Gas Industry

MARÉ Jean-Charles
Aerospace Actuators 3: European Commercial Aircraft and Tiltrotor Aircraft

MAXA Jean-Aimé, BEN MAHMOUD Mohamed Slim, LARRIEU Nicolas
Model-driven Development for Embedded Software: Application to Communications for Drone Swarm

MBIHI Jean
Analog Automation and Digital Feedback Control Techniques
Advanced Techniques and Technology of Computer-Aided Feedback Control

MORANA Joëlle
Logistics

SIMON Christophe, WEBER Philippe, SALLAK Mohamed
Data Uncertainty and Important Measures
(Systems Dependability Assessment Set – Volume 3)

TANIGUCHI Eiichi, THOMPSON Russell G.
City Logistics 1: New Opportunities and Challenges
City Logistics 2: Modeling and Planning Initiatives
City Logistics 3: Towards Sustainable and Liveable Cities

ZELM Martin, JAEKEL Frank-Walter, DOUMEINGTS Guy, WOLLSCHLAEGER Martin
Enterprise Interoperability: Smart Services and Business Impact of Enterprise Interoperability

2017

ANDRÉ Jean-Claude
From Additive Manufacturing to 3D/4D Printing 1: From Concepts to Achievements
From Additive Manufacturing to 3D/4D Printing 2: Current Techniques, Improvements and their Limitations
From Additive Manufacturing to 3D/4D Printing 3: Breakthrough Innovations: Programmable Material, 4D Printing and Bio-printing

ARCHIMÈDE Bernard, VALLESPIR Bruno
Enterprise Interoperability: INTEROP-PGSO Vision

CAMMAN Christelle, FIORE Claude, LIVOLSI Laurent, QUERRO Pascal
Supply Chain Management and Business Performance: The VASC Model

FEYEL Philippe
Robust Control, Optimization with Metaheuristics

MARÉ Jean-Charles
Aerospace Actuators 2: Signal-by-Wire and Power-by-Wire

POPESCU Dumitru, AMIRA Gharbi, STEFANOIU Dan, BORNE Pierre
Process Control Design for Industrial Applications

RÉVEILLAC Jean-Michel
Modeling and Simulation of Logistics Flows 1: Theory and Fundamentals
Modeling and Simulation of Logistics Flows 2: Dashboards, Traffic Planning and Management
Modeling and Simulation of Logistics Flows 3: Discrete and Continuous Flows in 2D/3D

2016

ANDRÉ Michel, SAMARAS Zissis
Energy and Environment
(Research for Innovative Transports Set - Volume 1)

AUBRY Jean-François, BRINZEI Nicolae, MAZOUNI Mohammed-Habib
Systems Dependability Assessment: Benefits of Petri Net Models (Systems Dependability Assessment Set - Volume 1)

BLANQUART Corinne, CLAUSEN Uwe, JACOB Bernard
Towards Innovative Freight and Logistics (Research for Innovative Transports Set - Volume 2)

COHEN Simon, YANNIS George
Traffic Management (Research for Innovative Transports Set - Volume 3)

MARÉ Jean-Charles
Aerospace Actuators 1: Needs, Reliability and Hydraulic Power Solutions

REZG Nidhal, HAJEJ Zied, BOSCHIAN-CAMPANER Valerio
Production and Maintenance Optimization Problems: Logistic Constraints and Leasing Warranty Services

TORRENTI Jean-Michel, LA TORRE Francesca
Materials and Infrastructures 1 (Research for Innovative Transports Set - Volume 5A)
Materials and Infrastructures 2 (Research for Innovative Transports Set - Volume 5B)

WEBER Philippe, SIMON Christophe
Benefits of Bayesian Network Models
(Systems Dependability Assessment Set – Volume 2)

YANNIS George, COHEN Simon
Traffic Safety (Research for Innovative Transports Set - Volume 4)

2015

AUBRY Jean-François, BRINZEI Nicolae
Systems Dependability Assessment: Modeling with Graphs and Finite State Automata

BOULANGER Jean-Louis
CENELEC 50128 and IEC 62279 Standards

BRIFFAUT Jean-Pierre
E-Enabled Operations Management

MISSIKOFF Michele, CANDUCCI Massimo, MAIDEN Neil
Enterprise Innovation

2014

CHETTO Maryline
Real-time Systems Scheduling
Volume 1 – Fundamentals
Volume 2 – Focuses

DAVIM J. Paulo
Machinability of Advanced Materials

ESTAMPE Dominique
Supply Chain Performance and Evaluation Models

FAVRE Bernard
Introduction to Sustainable Transports

GAUTHIER Michaël, ANDREFF Nicolas, DOMBRE Etienne
Intracorporeal Robotics: From Milliscale to Nanoscale

MICOUIN Patrice
Model Based Systems Engineering: Fundamentals and Methods

MILLOT Patrick
Designing Human–Machine Cooperation Systems

NI Zhenjiang, PACORET Céline, BENOSMAN Ryad, RÉGNIER Stéphane
Haptic Feedback Teleoperation of Optical Tweezers

OUSTALOUP Alain
Diversity and Non-integer Differentiation for System Dynamics

REZG Nidhal, DELLAGI Sofien, KHATAD Abdelhakim
Joint Optimization of Maintenance and Production Policies

STEFANOIU Dan, BORNE Pierre, POPESCU Dumitru, FILIP Florin Gh., EL KAMEL Abdelkader
Optimization in Engineering Sciences: Metaheuristics, Stochastic Methods and Decision Support

2013

ALAZARD Daniel
Reverse Engineering in Control Design

ARIOUI Hichem, NEHAOUA Lamri
Driving Simulation

CHADLI Mohammed, COPPIER Hervé
Command-control for Real-time Systems

DAAFOUZ Jamal, TARBOURIECH Sophie, SIGALOTTI Mario
Hybrid Systems with Constraints

FEYEL Philippe
Loop-shaping Robust Control

FLAUS Jean-Marie
Risk Analysis: Socio-technical and Industrial Systems

FRIBOURG Laurent, SOULAT Romain
Control of Switching Systems by Invariance Analysis: Application to Power Electronics

GROSSARD Mathieu, REGNIER Stéphane, CHAILLET Nicolas
Flexible Robotics: Applications to Multiscale Manipulations

GRUNN Emmanuel, PHAM Anh Tuan
Modeling of Complex Systems: Application to Aeronautical Dynamics

HABIB Maki K., DAVIM J. Paulo
Interdisciplinary Mechatronics: Engineering Science and Research Development

HAMMADI Slim, KSOURI Mekki
Multimodal Transport Systems

JARBOUI Bassem, SIARRY Patrick, TEGHEM Jacques
Metaheuristics for Production Scheduling

KIRILLOV Oleg N., PELINOVSKY Dmitry E.
Nonlinear Physical Systems

LE Vu Tuan Hieu, STOICA Cristina, ALAMO Teodoro,
CAMACHO Eduardo F., DUMUR Didier
Zonotopes: From Guaranteed State-estimation to Control

MACHADO Carolina, DAVIM J. Paulo
Management and Engineering Innovation

MORANA Joëlle
Sustainable Supply Chain Management

SANDOU Guillaume
Metaheuristic Optimization for the Design of Automatic Control Laws

STOICAN Florin, OLARU Sorin
Set-theoretic Fault Detection in Multisensor Systems

2012

AÏT-KADI Daoud, CHOUINARD Marc, MARCOTTE Suzanne, RIOPEL Diane
Sustainable Reverse Logistics Network: Engineering and Management

BORNE Pierre, POPESCU Dumitru, FILIP Florin G., STEFANOIU Dan
Optimization in Engineering Sciences: Exact Methods

CHADLI Mohammed, BORNE Pierre
Multiple Models Approach in Automation: Takagi-Sugeno Fuzzy Systems

DAVIM J. Paulo
Lasers in Manufacturing

DECLERCK Philippe
Discrete Event Systems in Dioid Algebra and Conventional Algebra

DOUMIATI Moustapha, CHARARA Ali, VICTORINO Alessandro,
LECHNER Daniel
Vehicle Dynamics Estimation using Kalman Filtering: Experimental Validation

GUERRERO José A, LOZANO Rogelio
Flight Formation Control

HAMMADI Slim, KSOURI Mekki
Advanced Mobility and Transport Engineering

MAILLARD Pierre
Competitive Quality Strategies

MATTA Nada, VANDENBOOMGAERDE Yves, ARLAT Jean
Supervision and Safety of Complex Systems

POLER Raul *et al.*
Intelligent Non-hierarchical Manufacturing Networks

TROCCAZ Jocelyne
Medical Robotics

YALAOUI Alice, CHEHADE Hicham, YALAOUI Farouk, AMODEO Lionel
Optimization of Logistics

ZELM Martin *et al.*
Enterprise Interoperability –I-EASA12 Proceedings

2011

CANTOT Pascal, LUZEAUX Dominique
Simulation and Modeling of Systems of Systems

DAVIM J. Paulo
Mechatronics

DAVIM J. Paulo
Wood Machining

GROUS Ammar
Applied Metrology for Manufacturing Engineering

KOLSKI Christophe
Human–Computer Interactions in Transport

LUZEAUX Dominique, RUAULT Jean-René, WIPPLER Jean-Luc
Complex Systems and Systems of Systems Engineering

ZELM Martin, *et al.*
Enterprise Interoperability: IWEI2011 Proceedings

2010

BOTTA-GENOULAZ Valérie, CAMPAGNE Jean-Pierre, LLERENA Daniel, PELLEGRIN Claude
Supply Chain Performance / Collaboration, Alignement and Coordination

BOURLÈS Henri, GODFREY K.C. Kwan
Linear Systems

BOURRIÈRES Jean-Paul
Proceedings of CEISIE'09

CHAILLET Nicolas, REGNIER Stéphane
Microrobotics for Micromanipulation

DAVIM J. Paulo
Sustainable Manufacturing

GIORDANO Max, MATHIEU Luc, VILLENEUVE François
Product Life-Cycle Management / Geometric Variations

LOZANO Rogelio
Unmanned Aerial Vehicles / Embedded Control

LUZEAUX Dominique, RUAULT Jean-René
Systems of Systems

VILLENEUVE François, MATHIEU Luc
Geometric Tolerancing of Products

2009

DIAZ Michel
Petri Nets / Fundamental Models, Verification and Applications

OZEL Tugrul, DAVIM J. Paulo
Intelligent Machining

PITRAT Jacques
Artificial Beings

2008

ARTIGUES Christian, DEMASSEY Sophie, NÉRON Emmanuel
Resources–Constrained Project Scheduling

BILLAUT Jean-Charles, MOUKRIM Aziz, SANLAVILLE Eric
Flexibility and Robustness in Scheduling

DOCHAIN Denis
Bioprocess Control

LOPEZ Pierre, ROUBELLAT François
Production Scheduling

THIERRY Caroline, THOMAS André, BEL Gérard
Supply Chain Simulation and Management

2007

DE LARMINAT Philippe
Analysis and Control of Linear Systems

DOMBRE Etienne, KHALIL Wisama
Robot Manipulators

LAMNABHI Françoise *et al.*
Taming Heterogeneity and Complexity of Embedded Control

LIMNIOS Nikolaos
Fault Trees

2006

FRENCH COLLEGE OF METROLOGY
Metrology in Industry

NAJIM Kaddour
Control of Continuous Linear Systems